MR TOMPKINS IN PAPERBACK

Mr Tompkins in Paperback

by

G. GAMOW

Illustrated by the author and

John Hookham

CAMBRIDGE UNIVERSITY PRESS

Cambridge
London New York New Rochelle
Melbourne Sydney

Published by the Press Syndicate of the University of Cambridge
The Pitt Building, Trumpington Street, Cambridge CB2 1RP
32 East 57th Street, New York, NY 10022, USA
10 Stamford Road, Oakleigh, Melbourne 3166, Australia

Mr Tompkins in Wonderland
First published 1940
Mr Tompkins explores the Atom
First published 1945

This Edition

First published 1965
Reprinted, with corrections 1967
Reprinted 1969, 1971, 1973, 1975, 1976, 1978,
1979, 1980, 1981, 1982, 1983, 1984, 1985, 1986

Printed in the United States of America

Library of Congress Catalogue
Card Number: 65-20791

ISBN: 0-521-09355-4 paperback

To my friend and editor
RONALD MANSBRIDGE

Preface

In the winter of 1938 I wrote a short, scientifically fantastic story (not a science fiction story) in which I tried to explain to the layman the basic ideas of the theory of curvature of space and the expanding universe. I decided to do this by exaggerating the actually existing relativistic phenomena to such an extent that they could easily be observed by the hero of the story, C. G. H.* Tompkins, a bank clerk interested in modern science.

I sent the manuscript to *Harper's Magazine* and, like all beginning authors, got it back with a rejection slip. The other half-a-dozen magazines which I tried followed suit. So I put the manuscript in a drawer of my desk and forgot about it. During the summer of the same year, I attended the International Conference of Theoretical Physics, organized by the League of Nations in Warsaw. I was chatting over a glass of excellent Polish miod with my old friend Sir Charles Darwin, the grandson of Charles (*The Origin of Species*) Darwin, and the conversation turned to the popularization of science. I told Darwin about the bad luck I had had along this line, and he said: 'Look, Gamow, when you get back to the United States dig up your manuscript and send it to Dr C. P. Snow, who is the editor of a popular scientific magazine *Discovery* published by the Cambridge University Press.'

So I did just this, and a week later came a telegram from Snow saying: 'Your article will be published in the next issue. Please send more.' Thus a number of stories on Mr Tompkins, which popularized the theory of relativity and the quantum theory,

* The initials of Mr Tompkins originated from three fundamental physical constants: the velocity of light c; the gravitational constant G; and the quantum constant h, which have to be changed by immensely large factors in order to make their effect easily noticeable by the man on the street.

appeared in subsequent issues of *Discovery*. Soon thereafter I received a letter from the Cambridge University Press, suggesting that these articles, with a few additional stories to increase the number of pages, should be published in book form. The book, called *Mr Tompkins in Wonderland*, was published by Cambridge University Press in 1940 and since that time has been reprinted sixteen times. This book was followed by the sequel, *Mr Tompkins Explores the Atom*, published in 1944 and by now reprinted nine times. In addition, both books have been translated into practically all European languages (except Russian), and also into Chinese and Hindi.

Recently the Cambridge University Press decided to unite the two original volumes into a single paperback edition, asking me to update the old material and add some more stories treating the advances in physics and related fields which took place after these books were originally published. Thus I had to add the stories on fission and fusion, the steady state universe, and exciting problems concerning elementary particles. This material forms the present book.

A few words must be said about the illustrations. The original articles in *Discovery* and the first original volume were illustrated by Mr John Hookham, who created the facial features of Mr Tompkins. When I wrote the second volume, Mr Hookham had retired from work as an illustrator, and I decided to illustrate the book myself, faithfully following Hookham's style. The new illustrations in the present volume are also mine. The verses and songs appearing in this volume are written by my wife Barbara.

G. GAMOW

University of Colorado,
Boulder, Colorado, U.S.A.

Contents

Acknowledgements

Thanks are due to the following for permission to reproduce copyright material: to Edward B. Marks Music Corporation for the settings of *O come, all ye Faithful* ('O Atome prreemorrdial', p. 57) and *Rule, Britannia* ('The Universe, by heavn's decree', p. 60) from *Time to Sing*; and to the Macmillan Company for figure A on p. 144 from *The Crystalline State*, by Sir W. H. Bragg and W. L. Bragg.

Introduction

From early childhood onwards we grow accustomed to the surrounding world as we perceive it through our five senses; in this stage of mental development the fundamental notions of space, time and motion are formed. Our mind soon becomes so accustomed to these notions that later on we are inclined to believe that our concept of the outside world based on them is the only possible one, and any idea of changing them seems paradoxical to us. However, the development of exact physical methods of observation and the profounder analysis of observed relations have brought modern science to the definite conclusion that this 'classical' foundation fails completely when used for the detailed description of phenomena ordinarily inaccessible to our everyday observation, and that, for the correct and consistent description of our new refined experience, some change in the fundamental concepts of space, time, and motion is absolutely necessary.

The deviations between the common notions and those introduced by modern physics are, however, negligibly small so far as the experience of ordinary life is concerned. If, however, we imagine other worlds, with the same physical laws as those of our own world, but with different numerical values for the physical constants determining the limits of applicability of the old concepts, the new and correct concepts of space, time and motion, at which modern science arrives only after very long and elaborate investigations, would become a matter of common knowledge. We may say that even a primitive savage in such a world would be acquainted with the principles of relativity and quantum theory, and would use them for his hunting purposes and everyday needs.

The hero of the present stories is transferred, in his dreams, into several worlds of this type, where the phenomena, usually inaccessible to our ordinary senses, are so strongly exaggerated that they could easily be observed as the events of ordinary life. He was helped in his fantastic but scientifically correct dream by an old professor of physics (whose daughter, Maud, he eventually married) who explained to him in simple language the unusual events which he observed in the world of relativity, cosmology, quantum, atomic and nuclear structure, elementary particles, etc.

It is hoped that the unusual experiences of Mr Tompkins will help the interested reader to form a clearer picture of the actual physical world in which we are living.

I

City Speed Limit

It was a bank holiday, and Mr Tompkins, the little clerk of a big city bank, slept late and had a leisurely breakfast. Trying to plan his day, he first thought about going to some afternoon movie and, opening the morning paper, turned to the entertainment page. But none of the films looked attractive to him. He detested all this Hollywood stuff, with infinite romances between popular stars.

All this Hollywood stuff!

If only there were at least one film with some real adventure, with something unusual and maybe even fantastic about it. But there was none. Unexpectedly, his eye fell on a little notice in the corner of the page. The local university was announcing a series of lectures on the problems of modern physics, and this afternoon's lecture was to be about EINSTEIN's Theory of Relativity. Well, that might be something! He had often heard the statement that only a dozen people in the world really understood Einstein's theory. Maybe he could become the thirteenth! Surely he would go to the lecture; it might be just what he needed.

He arrived at the big university auditorium after the lecture had begun. The room was full of students, mostly young, listening with keen attention to the tall, white-bearded man near the blackboard who was trying to explain to his audience the basic ideas of the Theory of Relativity. But Mr Tompkins got only as far as understanding that the whole point of Einstein's theory is that there is a maximum velocity, the velocity of light, which cannot be surpassed by any moving material body, and that this fact leads to very strange and unusual consequences. The professor stated, however, that as the velocity of light is 186,000 miles per second, the relativity effects could hardly be observed for events of ordinary life. But the nature of these unusual effects was really much more difficult to understand, and it seemed to Mr Tompkins that all this was contradictory to common sense. He was trying to imagine the contraction of measuring rods and the odd behaviour of clocks—effects which should be expected if they move with a velocity close to that of light—when his head slowly dropped on his shoulder.

When he opened his eyes again, he found himself sitting not on a lecture room bench but on one of the benches installed by the city for the convenience of passengers waiting for a bus. It was a beautiful old city with medieval college buildings lining the street. He suspected that he must be dreaming but to his surprise there was nothing unusual happening around him; even a policeman standing on the opposite corner looked as policemen usually do. The hands of the big clock on the tower down the street were pointing to five o'clock and the streets were nearly empty. A single cyclist was coming slowly down the street and, as he approached, Mr Tompkins's eyes opened wide with astonishment. For the bicycle and the young man on it were unbelievably shortened in the direction of the motion, as if seen through a cylindrical lens. The clock on the tower struck five, and the cyclist, evidently in a hurry, stepped harder on the pedals. Mr Tompkins did not notice that he gained much in speed, but, as

Unbelievably shortened

the result of his effort, he shortened still more and went down the street looking exactly like a picture cut out of cardboard. Then Mr Tompkins felt very proud because he could understand what was happening to the cyclist—it was simply the contraction of moving bodies, about which he had just heard. 'Evidently nature's speed limit is lower here,' he concluded, 'that is why the bobby on the corner looks so lazy, he need not watch for speeders.' In fact, a taxi moving along the street at the moment and making all the noise in the world could not do much better than the cyclist, and was just crawling along. Mr Tompkins decided to overtake the cyclist, who looked a good sort of fellow, and ask him all about it. Making sure that the policeman was looking the other way, he borrowed somebody's bicycle standing near the kerb and sped

The city blocks became still shorter

down the street. He expected that he would be immediately shortened, and was very happy about it as his increasing figure had lately caused him some anxiety. To his great surprise, however, nothing happened to him or to his cycle. On the other hand, the picture around him completely changed. The streets grew shorter, the windows of the shops began to look like narrow slits, and the policeman on the corner became the thinnest man he had ever seen.

'By Jove!' exclaimed Mr Tompkins excitedly, 'I see the trick now. This is where the word *relativity* comes in. Everything that moves relative to me looks shorter for me, whoever works the pedals!' He was a good cyclist and was doing his best to overtake the young man. But he found that it was not at all easy to get up speed on this bicycle. Although he was working on the pedals as hard as he possibly could, the increase in speed was almost negligible. His legs already began to ache, but still he could not manage to pass a lamp-post on the corner much faster than when he had just started. It looked as if all his efforts to move faster were leading to no result. He understood now very well why the cyclist and the cab he had just met could not do any better, and he remembered the words of the professor about the impossibility of surpassing the limiting velocity of light. He noticed, however, that the city blocks became still shorter and the cyclist riding ahead of him did not now look so far away. He overtook the cyclist at the second turning, and when they had been riding side by side for a moment, was surprised to see the cyclist was actually quite a normal, sporting-looking young man. 'Oh, that must be because we do not move relative to each other,' he concluded; and he addressed the young man.

'Excuse me, sir!' he said, 'Don't you find it inconvenient to live in a city with such a slow speed limit?'

'Speed limit?' returned the other in surprise, 'we don't have any speed limit here. I can get anywhere as fast as I wish, or at least I could if I had a motor-cycle instead of this nothing-to-be-done-with old bike!'

'But you were moving very slowly when you passed me a moment ago,' said Mr Tompkins. 'I noticed you particularly.'

'Oh you did, did you?' said the young man, evidently offended. 'I suppose you haven't noticed that since you first addressed me we have passed five blocks. Isn't that fast enough for you?'

'But the streets became so short,' argued Mr Tompkins.

'What difference does it make anyway, whether we move faster or whether the street becomes shorter? I have to go ten blocks to get to the post office, and if I step harder on the pedals the blocks become shorter and I get there quicker. In fact, here we are,' said the young man getting off his bike.

Mr Tompkins looked at the post office clock, which showed half-past five. 'Well!' he remarked triumphantly, 'it took you half an hour to go this ten blocks, anyhow—when I saw you first it was exactly five!'

'And did you *notice* this half hour?' asked his companion. Mr Tompkins had to agree that it had really seemed to him only a few minutes. Moreover, looking at his wrist watch he saw it was showing only five minutes past five. 'Oh!' he said, 'is the post office clock fast?' 'Of course it is, or your watch is too slow, just because you have been going too fast. What's the matter with you, anyway? Did you fall down from the moon?' and the young man went into the post office.

After this conversation, Mr Tompkins realized how unfortunate it was that the old professor was not at hand to explain all these strange events to him. The young man was evidently a native, and had been accustomed to this state of things even before he had learned to walk. So Mr Tompkins was forced to explore this strange world by himself. He put his watch right by the post office clock, and to make sure that it went all right waited for ten minutes. His watch did not lose. Continuing his journey down the street he finally saw the railway station and decided to check his watch again. To his surprise it was again quite a bit slow. 'Well, this must be some relativity effect, too,' concluded Mr Tompkins; and decided to ask about it from somebody more intelligent than the young cyclist.

The opportunity came very soon. A gentleman obviously in his forties got out of the train and began to move towards the exit. He was met by a very old lady, who, to Mr Tompkins's great surprise, addressed him as 'dear Grandfather'. This was too much for Mr Tompkins. Under the excuse of helping with the luggage, he started a conversation.

'Excuse me, if I am intruding into your family affairs,' said he, 'but are you really the grandfather of this nice old lady? You see, I am a stranger here, and I never....' 'Oh, I see,' said the gentleman, smiling with his moustache. 'I suppose you are taking me for the Wandering Jew or something. But the thing is really quite simple. My business requires me to travel quite a lot, and, as I spend most of my life in the train, I naturally grow old much more slowly than my relatives living in the city. I am so glad that I came back in time to see my dear little grand-daughter still alive! But excuse me, please, I have to attend to her in the taxi,' and he hurried away leaving Mr Tompkins alone again with his problems. A couple of sandwiches from the station buffet somewhat strengthened his mental ability, and he even went so far as to claim that he had found the contradiction in the famous principle of relativity.

'Yes, of course,' thought he, sipping his coffee, 'if all were relative, the traveller would appear to his relatives as a very old man, and they would appear very old to him, although both sides might in fact be fairly young. But what I am saying now is definitely nonsense: One could not have relative grey hair!' So he decided to make a last attempt to find out how things really are, and turned to a solitary man in railway uniform sitting in the buffet.

'Will you be so kind, sir,' he began, 'will you be good enough to tell me who is responsible for the fact that the passengers in the train grow old so much more slowly than the people staying at one place?'

'I am responsible for it,' said the man, very simply.

'Oh!' exclaimed Mr Tompkins. 'So you have solved the problem of the Philosopher's Stone of the ancient alchemists. You should be quite a famous man in the medical world. Do you occupy the chair of medicine here?'

'No,' answered the man, being quite taken aback by this, 'I am just a brakeman on this railway.'

'Brakeman! You mean a brakeman...,' exclaimed Mr Tompkins, losing all the ground under him. 'You mean you—just put the brakes on when the train comes to the station?'

'Yes, that's what I do: and every time the train gets slowed down, the passengers gain in their age relative to other people. Of course,' he added modestly, 'the engine driver who accelerates the train also does his part in the job.'

'But what has it to do with staying young?' asked Mr Tompkins in great surprise.

'Well, I don't know exactly,' said the brakeman, 'but it is so. When I asked a university professor travelling in my train once, how it comes about, he started a very long and incomprehensible speech about it, and finally said that it is something similar to 'gravitation redshift—I think he called it—on the sun. Have you heard anything about such things as redshifts?'

'No-o,' said Mr Tompkins, a little doubtfully; and the brakeman went away shaking his head.

Suddenly a heavy hand shook his shoulder, and Mr Tompkins found himself sitting not in the station café but in the chair of the auditorium in which he had been listening to the professor's lecture. The lights were dimmed and the room was empty. The janitor who wakened him said: 'We are closing up, Sir; if you want to sleep, better go home.' Mr Tompkins got to his feet and started toward the exit.

2

The Professor's Lecture on Relativity which caused Mr Tompkins's dream

Ladies and Gentlemen:

In a very primitive stage of development the human mind formed definite notions of space and time as the frame in which different events take place. These notions, without essential changes, have been carried forward from generation to generation, and, since the development of exact sciences, have been built into the foundations of the mathematical description of the universe. The great NEWTON perhaps gave the first clear-cut formulation of the classical notions of space and time, writing in his *Principia*:

'Absolute space, in its own nature, without relation to anything external, remains always similar and immovable;' and 'Absolute, true and mathematical time, of itself, and from its own nature, flows equably without relation to anything external.'

So strong was the belief in the absolute correctness of these classical ideas about space and time that they have often been held by philosophers as given *a priori*, and no scientist even thought about the possibility of doubting them.

However, just at the start of the present century it became clear that a number of results, obtained by most refined methods of experimental physics, led to clear contradictions if interpreted in the classical frame of space and time. This fact brought to one of the greatest contemporary physicists, ALBERT EINSTEIN, the revolutionary idea that there are hardly any reasons, except those of tradition, for considering the classical notions concerning space and time as absolutely true, and that they could and should be changed to fit our new and more refined experience. In fact, since the classical notions of space and time were formulated on the basis of human experience in ordinary life, we need not be surprised that the refined methods of observation of today, based on highly developed experimental technique, indicate that these old notions are too rough and inexact, and could have been used in ordinary life and in the earlier stages of development of physics only because their deviations from the correct notions were sufficiently small. Nor need we be surprised that the broadening of the field of exploration of modern science should bring us to regions where these deviations become so very large that the classical notions could not be used at all.

The most important experimental result which led to the fundamental criticism of our classical notions was *the discovery of the fact that the velocity of light in a vacuum represents the upper limit for all possible physical velocities.* This important and unexpected conclusion resulted mainly from the experiments of the American physicist, MICHELSON, who, at the end of last century, tried to observe the effect of the motion of the earth on the velocity of propagation of light and, to his great surprise and the surprise of all the scientific world, found that no such effect exists and that the velocity of light in a vacuum comes out always exactly the same independent of the system from which it is measured or the motion of the source from which it is emitted. There is no need to explain that such a result is extremely unusual and contradicts our most fundamental concepts concerning motion. In fact, if something is moving fast through space and you yourself move so as to meet it, the moving object will strike you with greater relative velocity, equal to the sum of velocity of the object and the observer. On the other hand, if you run away from it, it will hit you from behind with smaller velocity, equal to the difference of the two velocities.

Also, if you move, say in a car, to meet the sound propagating through the air, the velocity of the sound as measured in the car will be larger by the amount of your driving speed, or it will be correspondingly small if the sound is overtaking you. We call it the *theorem of addition of velocities* and it was always held to be self-evident.

However, the most careful experiments have shown that, in the case of light, it is no longer true, the velocity of light in a vacuum remaining always the same and equal to 300,000 km per second (we usually denote it by the symbol c), independent of how fast the observer himself is moving.

'Yes,' you will say, 'but is it not possible to construct a super-light velocity by adding several smaller velocities which can be physically attained?'

For example, we could consider a very fast-moving train, say, with three quarters the velocity of light and a tramp running along the roofs of the carriages also with three-quarters of the velocity of light.

According to the theorem of addition the total velocity should be one and a half times that of light, and the running tramp should be able to overtake the beam of light from a signal lamp. The truth, however, is that, since the constancy of the velocity of light is an experimental fact, the resulting velocity in our case must be smaller than we expect—it cannot surpass the critical value c; and thus we come to the conclusion that, for smaller velocities also, the classical theorem of addition must be wrong.

The mathematical treatment of the problem, into which I do not want to enter here, leads to a very simple new formula for the calculation of the resulting velocity of two superimposed motions.

If v_1 and v_2 are the two velocities to be added, the resulting velocity comes out to be

$$V = \frac{v_1 \pm v_2}{1 \pm \frac{v_1 v_2}{c^2}}. \qquad (1)$$

You see from this formula that if both original velocities were small, I mean small as compared with the velocity of light, the second term in the denominator of (1) can be neglected as compared with unity and you have the classical theorem of addition of velocities. If, however, v_1 and v_2 are not small the result will be always somewhat smaller than the arithmetical sum. For instance, in the example of our tramp running along a train, $v_1 = \frac{3}{4}c$ and $v_2 = \frac{3}{4}c$, and our formula gives for the resulting velocity $V = \frac{24}{25}c$, which is still smaller than the velocity of light.

In a particular case, when one of the original velocities is c, formula (1) gives c for the resulting velocity independent of what the second velocity may be. Thus, by overlapping any number of velocities, we can never surpass the velocity of light.

You might also be interested to know that this formula has been proved experimentally and it was really found that the resultant of two velocities is always somewhat smaller than their arithmetical sum.

Recognizing the existence of the upper-limit velocity we can start on the criticism of the classical ideas of space and time, directing our first blow against the notion of *simultaneousness* based upon them.

When you say, 'The explosion in the mines near Capetown happened at exactly the same moment as the ham and eggs were being served in your London apartment,' you think you know what you mean. I am going to show you, however, that you do not, and that, strictly speaking, this statement has no exact meaning. In fact, what method would you use to check whether two events in two different places are simultaneous or not? You would say that the clock at both places would show the same time; but then the question arises how to set the distant clocks so that they would show the same time simultaneously, and we are back at the original question.

Since the independence of the velocity of light in a vacuum on the motion of its source or the system in which it is measured is one of the most exactly established experimental facts, the following method of measuring the distances and setting the clock correctly on different observational stations should be recognized as the most rational and, as you will agree after thinking more about it, the only reasonable method.

A light signal is sent from the station *A*, and as soon as it is received at the station *B* it is returned back to *A*. One-half of the time, as read at station *A*, between the sending and the return of the signal, multiplied by the constant velocity of light, will be defined as the distance between *A* and *B*.

The clocks on stations *A* and *B* are said to be set correctly if at the moment of arrival of the signal at *B* the local clock were showing just the average of two times recorded at *A* at the moments of sending and receiving the signal. Using this method

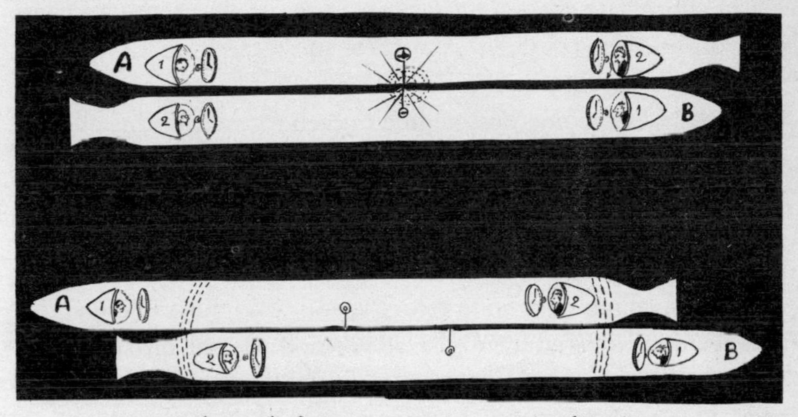

Two long platforms moving in opposite directions

between different observational stations established on a rigid body we arrive finally at the desired frame of reference, and can answer questions concerning the simultaneousness or time interval between two events in different places.

But will these results be recognized by observers on the other systems? To answer this question, let us suppose that such frames of reference have been established on two different rigid bodies, say on two long space rockets moving with a constant speed in opposite directions, and let us now see how these two frames will check with one another. Suppose four observers are located on the front- and the rear-ends of each rocket and want first of all to set their clocks correctly. Each pair of observers can use on their rockets the modification of the above-mentioned method by sending a light signal from the middle of the rocket (as measured by measuring-stick) and setting zero point on their watches when the signal, coming from the middle of the rocket, arrives at each end of it. Thus, each pair of our observers has established, according to previous definition, the criterion of simultaneousness in their own system and have set their watches 'correctly' from their point of view, of course.

Now they decide to see whether the time readings on their rocket check with that on the other. For example, do the watches of two observers on different rockets show the same time when they are passing one another? This can be tested by the following method: In the geometrical middle of each rocket they install two electrically charged conductors, in such a way that, when the rockets pass each other, a spark jumps between the conductors, and light signals start simultaneously from the centre of each platform towards its front and rear ends. By the time the light signals, travelling with finite velocity, approach the observers, the rockets have changed their relative position and the observers 2*A* and 2*B* will be closer to the source of light than the observers 1*A* and 1*B*.

It is clear that when the light signal reaches the observer 2*A*, the observer 1*B* will be farther behind, so that the signal will take some additional time to reach him. Thus, if the watch of 1*B* is set in such a way as to show zero time at the arrival of the signal, the observer 2*A* will insist that it is behind the correct time.

In the same way another observer, 1*A*, will come to the conclusion that the watch of 2*B*, who met the signal before him, is ahead of time. Since, according to their definition of simultaneousness, their own watches are set correctly, the observers on rocket *A* will agree that there is a difference between the watches of the observers on rocket *B*. We should not, however, forget that the observers on rocket *B*, for exactly the same reasons, will consider their own watches as set correctly but will claim that a difference of setting exists between the watches on rocket *A*.

Since both rockets are quite equivalent, this quarrel between the two groups of observers can be settled only by saying that both groups are correct from their own point of view, but that the question who is correct 'absolutely' has no physical sense.

I am afraid I have made you quite tired by these long considerations, but if you follow them carefully it will be clear to you that, as soon as our method of space–time measurement is adopted, *the notion of absolute simultaneousness vanishes, and two events in different places considered as simultaneous from one system of reference will be separated by a definite time interval from the point of view of another system.*

This proposition sounds at first extremely unusual, but does it look unusual to you if I say that, having your dinner on a train, you eat your soup and your dessert in the same point of the dining car, but in widely separated points of the railway track? However, this statement about your dinner in the train can be formulated by saying that *two events happening at different times at the same point of one system of reference will be separated by a definite space interval from the point of view of another system.*

If you compare this 'trivial' proposition with the previous 'paradoxical' one, you will see that they are absolutely symmetrical and can be transformed into one another simply by exchanging the words 'time' and 'space'.

Here is the whole point of Einstein's view: whereas in classical physics time was considered as something quite independent of space and motion 'flowing equably without relation to anything external' (Newton), in the new physics space and time are closely connected and represent just two different cross-sections of one homogeneous 'space–time continuum', in which all observable events take place. The splitting of this four-dimensional continuum into three-dimensional space and one-dimensional time is purely arbitrary, and depends on the system from which the observations are made.

Two events, separated in space by the distance l and in time by the interval t as observed in one system, will be separated by another distance l' and another time interval t' as seen from another system, so that, in a certain sense one can speak about the transformation of space into time and vice versa. It is also not difficult to see why the transformation of time into space, as in the example of the dinner in a train, is quite a common notion for us, whereas the transformation of space into time, resulting in the relativity of simultaneousness, seems very unusual. The point is that if we measure distances, say, in 'centimetres', the corresponding unit of time should be not the conventional 'second' but a 'rational unit of time', represented by the interval of time necessary for a light signal to cover a distance of one centimetre, i.e. 0·000,000,000,03 second.

Therefore, in the sphere of our ordinary experience the transformation of space intervals into time intervals leads to results practically unobservable, which seems to support the classical view that time is something absolutely independent and unchangeable.

However, when investigating motions with very high velocities, as, for example, the motion of electrons thrown out from radioactive bodies or the motion of electrons inside an atom, where the distances covered in a certain interval of time are of the same order of magnitude as the time expressed in rational units, one necessarily meets with both of the effects discussed above and the theory of relativity becomes of great importance. Even in the region of comparatively small velocities, as, for example, the motion of planets in our solar system, relativistic effects can be observed owing to the extreme precision of astronomical measurements; such observation of relativistic effects requires, however, measurements of the changes of planetary motion amounting to a fraction of an angular second per year.

As I have tried to explain to you, the criticism of the notions of space and time leads to the conclusion that space intervals can be partially converted into time intervals and the other way round; which means that the numerical value of a given distance or period of time will be different as measured from different moving systems.

A comparatively simple mathematical analysis of this problem, into which I do not, however, want to enter in these lectures, leads to a definite formula for the change of these values. It works out that any object of length l, moving relative to the observer with velocity v, will be shortened by an amount depending on its velocity, and its measured length will be

$$l' = l\sqrt{1-\frac{v^2}{c^2}}. \tag{2}$$

Analogously, any process taking time t will be observed from the relatively moving system as taking a longer time t', given by

$$t' = \frac{t}{\sqrt{1-\frac{v^2}{c^2}}}. \tag{3}$$

This is the famous 'shortening of space' and 'expanding of time' in the theory of relativity.

Ordinarily, when v is very much less than c the effects are very small, but, for sufficiently large velocities, the lengths as observed from a moving system may be made arbitrarily small and time intervals arbitrarily long.

I do not want you to forget that both these effects are absolutely symmetrical systems, and, whereas the passengers on a fast-moving train will wonder why the people on the standing train are so lean and move so slowly, the passengers on the standing train will think the same about the people on the moving one.

Another important consequence of the existence of the maximum possible velocity pertains to the *mass* of moving bodies.

According to the general foundation of mechanics, the mass of a body determines the difficulty of setting it into motion or accelerating the motion already existing; the larger the mass, the more difficult it is to increase the velocity by a given amount.

The fact that no body under any circumstances can exceed the velocity of light leads us directly to the conclusion that its resistance to further acceleration or, in other words, its mass, must increase without limit when its velocity approaches the velocity of light. Mathematical analysis leads to a formula for this dependence, which is analogous to the formulae (2) and (3). If m_0 is the mass for very small velocities, the mass m at the velocity v is given by

$$m = \frac{m_0}{\sqrt{1 - \frac{v^2}{c^2}}} \tag{4}$$

and the resistance to further acceleration becomes infinite when v approaches c.

This effect of the relativistic change of mass can be easily observed experimentally on very fast-moving particles. For example, the mass of electrons emitted by radioactive bodies (with a velocity of 99% of that of light) is several times larger than in a state of rest and the masses of electrons forming so-called cosmic rays and moving often with 99·98% of the velocity of light are 1000 times larger. For such velocities the classical mechanics becomes absolutely inapplicable and we enter into the domain of the pure theory of relativity.

3

Mr Tompkins takes a holiday

Mr Tompkins was very amused about his adventures in the relativistic city, but was sorry that the professor had not been with him to give any explanation of the strange things he had observed: the mystery of how the railway brakeman had been able to prevent the passengers from getting old worried him especially. Many a night he went to bed with the hope that he would see this interesting city again, but the dreams were rare and mostly unpleasant; last time it was the manager of the bank who was firing him for the uncertainty he introduced into the bank accounts... so now he decided that he had better take a holiday, and go for a week somewhere to the sea. Thus he found himself sitting in a compartment of a train and watching through the window the grey roofs of the city suburb gradually giving place to the green meadows of the countryside. He picked up a newspaper and tried to interest himself in the Vietnam conflict. But it all seemed to be so dull, and the railway carriage rocked him pleasantly....

When he lowered the paper and looked out of the window again the landscape had changed considerably. The telegraph poles were so close to each other that they looked like a hedge, and the trees had extremely narrow crowns and were like Italian cypresses. Opposite to him sat his old friend the professor, looking through the window with great interest. He had probably got in while Mr Tompkins was busy with his newspaper.

'We are in the land of relativity,' said Mr Tompkins, 'aren't we?'

'Oh!' exclaimed the professor, 'you know so much already! Where did you learn it from?'

'I have already been here once, but did not have the pleasure of your company then.'

'So you are probably going to be my guide this time,' the old man said.

'I should say not,' retorted Mr Tompkins. 'I saw a lot of unusual things, but the local people to whom I spoke could not understand what my trouble was at all.'

'Naturally enough,' said the professor. 'They are born in this world and consider all the phenomena happening around them as self-evident. But I imagine they would be quite surprised if they happened to get into the world in which you used to live. It would look so remarkable to them.'

'May I ask you a question?' said Mr Tompkins. 'Last time I was here, I met a brakeman from the railway who insisted that owing to the fact that the train stops and starts again the passengers grow old less quickly than the people in the city. Is this magic, or is it also consistent with modern science?'

'There is never any excuse for putting forward magic as an explanation,' said the professor. 'This follows directly from the laws of physics. It was shown by Einstein, on the basis of his analysis of new (or should I say as-old-as-the-world but newly discovered) notions of space and time, that all physical processes slow down when the system in which they are taking place is changing its velocity. In our world the effects are almost unobservably small, but here, owing to the small velocity of light, they are usually very obvious. If, for example, you tried to boil an egg here, and instead of letting the saucepan stand quietly on the stove moved it to and fro, constantly changing its velocity, it would take you not five but perhaps six minutes to boil it properly. Also in the human body all processes slow down, if the person is sitting (for example) in a rocking chair or in a train which changes its speed; we live more slowly under such conditions. As, however, all processes slow down to the same extent, physicists prefer to say that *in a non-uniformly moving system time flows more slowly*.'

'But do scientists actually observe such phenomena in our world at home?'

'They do, but it requires considerable skill. It is technically very difficult to get the necessary accelerations, but the conditions existing in a non-uniformly moving system are analogous, or should I say identical, to the result of the action of a very large force of gravity. You may have noticed that when you are in an elevator which is rapidly accelerated upwards it seems to you that you have grown heavier; on the contrary, if the elevator starts downward (you realize it best when the rope breaks) you feel as though you were losing weight. The explanation is that the gravitational field created by acceleration is added to or subtracted from the gravity of the earth. Well, the potential of gravity on the sun is much larger than on the surface of the earth and all processes there should be therefore slightly slowed down. Astronomers do observe this.'

'But they cannot go to the sun to observe it?'

'They do not need to go there. They observe the light coming to us from the sun. This light is emitted by the vibration of different atoms in the solar atmosphere. If all processes go slower there, the speed of atomic vibrations also decreases, and by comparing the light emitted by solar and terrestrial sources one can see the difference. Do you know, by the way'—the professor interrupted himself—'what the name of this little station is that we are now passing?'

The train was rolling along the platform of a little countryside station which was quite empty except for the station master and a young porter sitting on a luggage trolley and reading a newspaper. Suddenly the station master threw his hands into the air and fell down on his face. Mr Tompkins did not hear the sound of shooting, which was probably lost in the noise of the train, but the pool of blood forming round the body of the station master left no doubt. The professor immediately pulled the emergency cord and the train stopped with a jerk. When they got out of the carriage the young porter was running towards the body, and a country policeman was approaching.

'Shot through the heart,' said the policeman after inspecting the body, and, putting a heavy hand on the porter's shoulder, he went on: 'I am arresting you for the murder of the station master.'

'I didn't kill him,' exclaimed the unfortunate porter. 'I was reading a newspaper when I heard the shot. These gentlemen from the train have probably seen all and can testify that I am innocent.'

'Yes,' said Mr Tompkins, 'I saw with my own eyes that this man was reading his paper when the station master was shot. I can swear it on the Bible.'

'But you were in the moving train,' said the policeman, taking an authoritative tone, 'and what you saw is therefore no evidence at all. As seen from the platform the man could have been shooting at the very same moment. Don't you know that simultaneousness depends on the system from which you observe it? Come along quietly,' he said, turning to the porter.

'Excuse me, constable,' interrupted the professor, 'but you are absolutely wrong, and I do not think that at headquarters they will like your ignorance. It is true, of course, that the notion of simultaneousness is highly relative in your country. It is also true that two events in different places could be simultaneous or not, depending on the motion of the observer. But, even in your country, no observer could see the consequence before the cause. You have never received a telegram before it was sent, have you? or got drunk before opening the bottle? As I understand you, you suppose that owing to the motion of the train the shooting would have been seen by us much *later* than its effect and, as we got out of the train immediately we saw the station master fall, we still had not seen the shooting itself. I know that in the police force you are taught to believe only what is written in your instructions, but look into them and probably you will find something about it.'

The professor's tone made quite an impression on the policeman and, pulling out his pocket book of instructions, he started to read it slowly through. Soon a smile of embarrassment spread out across his big, red face.

'Here it is,' said he, 'section 37, subsection 12, paragraph *e*: "As a perfect alibi should be recognized any authoritative proof, from any moving system whatsoever, that at the moment of the crime or within a time interval $\pm cd$ (c being natural speed limit and d the distance from the place of the crime) the suspect was seen in another place."'

'You are free, my good man,' he said to the porter, and then, turning to the professor: 'Thank you very much, Sir, for saving me from trouble with headquarters. I am new to the force and not yet accustomed to all these rules. But I must report the murder anyway,' and he went to the telephone box. A minute later he was shouting across the platform. 'All is in order now! They caught the real murderer when he was running away from the station. Thank you once more!'

'I may be very stupid,' said Mr Tompkins, when the train started again, 'but what is all this business about simultaneousness? Has it really no meaning in this country?'

'It has,' was the answer, 'but only to a certain extent; otherwise I should not have been able to help the porter at all. You see, the existence of a natural speed limit for the motion of any body or the propagation of any signal, makes simultaneousness in our ordinary sense of the word lose its meaning. You probably will see it more easily this way. Suppose you have a friend living in a far-away town, with whom you correspond by letter, mail train being the fastest means of communication. Suppose now that something happens to you on Sunday and you learn that the same thing is going to happen to your friend. It is clear that you cannot let him know about it before Wednesday. On the other hand, if he knew in advance about the thing that was going to happen to you, the last date to let you know about it would have been the previous Thursday. Thus for six days, from Thursday to next Wednesday, your friend was not able either to influence your fate on Sunday or to learn about it. From the point of view of causality he was, so to speak, excommunicated from you for six days.'

'What about a telegram?' suggested Mr Tompkins.

'Well, I accepted that the velocity of the mail train was the maximum possible velocity, which is about correct in this country. At home the velocity of light is the maximum velocity and you cannot send a signal faster than by radio.'

'But still,' said Mr Tompkins, 'even if the velocity of the mail train could not be surpassed, what has it to do with simultaneousness? My friend and myself would still have our Sunday dinners simultaneously, wouldn't we?'

'No, that statement would not have any sense then; one observer would agree to it, but there would be others, making their observations from different trains, who would insist that you eat your Sunday dinner at the same time as your friend has his Friday breakfast or Tuesday lunch. But in no way could anybody observe you and your friend simultaneously having meals more than three days apart.'

'But how can all this happen?' exclaimed Mr Tompkins unbelievingly.

'In a very simple way, as you might have noticed from my lectures. The upper limit of velocity must remain the same as observed from different moving systems. If we accept this we should conclude that....'

But their conversation was interrupted by the train arriving at the station at which Mr Tompkins had to get out.

When Mr Tompkins came down to have his breakfast in the long glass verandah of the hotel, the morning after his arrival at the seaside, a great surprise awaited him. At the table in the opposite corner sat the old professor and a pretty girl who was cheerfully relating something to the old man, and glancing often in the direction of the table where Mr Tompkins was sitting.

'I suppose I did look very stupid, sleeping in that train,' thought Mr Tompkins, getting more and more angry with himself. 'And the professor probably still remembers the stupid question I asked him about getting younger. But this at least will give me an opportunity to become better acquainted with him now and ask about the things I still do not understand.' He did not want to admit even to himself that it was not only conversation with the professor he was thinking about.

'Oh, yes, yes, I think I do remember seeing you at my lectures,' said the professor when they were leaving the dining room. 'This is my daughter, Maud. She is studying painting.'

'Very happy to meet you, Miss Maud,' said Mr Tompkins, and thought that this was the most beautiful name he had ever heard. 'I expect these surroundings must give you wonderful material for your sketches.'

'She will show them to you some time,' said the professor, 'but tell me, did you gather much from listening to my lecture?'

'Oh yes, I did, quite a lot—and in fact I myself experienced all these relativistic contractions of material objects and the crazy behaviour of clocks when I visited a city where the velocity of light was only about ten miles per hour.'

'Then it is a pity,' said the professor, 'that you missed my following lecture about the curvature of space and its relation to the forces of Newtonian gravity. But here on the beach we will have time, so that I will be able to explain all that to you. Do you, for example, understand the difference between the positive and negative curvature of space?'

'Daddy,' said Miss Maud, pouting her lips, 'if you are talking physics again, I think I will go and do some work.'

'All right, girlie, you run along,' said the professor, plunging himself into an easy chair. 'I see you did not study mathematics much, young man; but I think I can explain it to you very simply, taking, for simplicity, the example of a surface. Imagine that Mr Shell—you know, the man who owns the petrol stations—decides to see whether his stations are distributed uniformly throughout some country, say America. To do this, he gives orders to his office, somewhere in the middle of the country

(Kansas City is, I believe, considered as the heart of America), to count the number of stations within one hundred, two hundred, three hundred and so on miles from the city. He remembers from his school days that the area of a circle is proportional to the square of its radius, and expects that in the case of uniform distribution the number of stations thus counted should increase like the sequence of numbers 1; 4; 9; 16 and so on. When the report

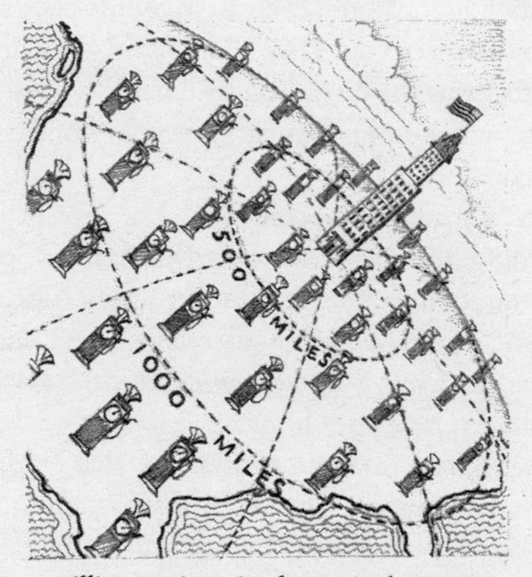

Filling stations in the United States

comes in, he will be very much surprised to see that the actual number of stations is increasing much more slowly, going, let us say, 1; 3·8; 8·5; 15·0; and so on. "What a mess," he would exclaim; "my managers in America do not know their job. What is the great idea of concentrating the stations near Kansas City?" But is he right in this conclusion?'

'Is he?' repeated Mr Tompkins, who was thinking about something else.

'He is not,' said the professor gravely. 'He has forgotten that the earth's surface is not a plane but a sphere. And on a sphere the area within a given radius grows more slowly with the radius than on a plane. Can't you really see it? Well, take a globe and try to see it for yourself. If, for example, you are on the north pole, the circle with the radius equal to a half meridian is the equator, and the area included is the northern hemisphere. Increase the radius twice and you will get in all the earth's surface; the area will increase only twice instead of four times as it would on a plane. Isn't it clear to you now?'

'It is,' said Mr Tompkins, making an effort to be attentive. 'And is this a positive or a negative curvature?'

'It is called positive curvature, and, as you see from the example of the globe, it corresponds to a finite surface having definite area. An example of a surface with negative curvature is given by a saddle.'

'By a saddle?' repeated Mr Tompkins.

'Yes, by a saddle, or, on the surface of the earth, by a saddle pass between two mountains. Suppose a botanist lives in a mountain hut situated on such a saddle pass and is interested in the density of growth of pines around the hut. If he counts the number of pines growing within one hundred, two hundred, and so on feet from the hut, he will find that the number of pines increases faster than the square of the distance, the point being that on a saddle surface the area included within a given radius is larger than on a plane. Such surfaces are said to possess a negative curvature. If you try to spread a saddle surface on a plane you will have to make folds in it, whereas doing the same with a spherical surface you will probably tear it if it is not elastic.'

'I see,' said Mr Tompkins. 'And you mean to say that a saddle surface is infinite although curved.'

'Exactly so,' approved the professor. 'A saddle surface extends to infinity in all directions and never closes on itself. Of course, in my example of a saddle pass the surface ceases to possess negative curvature as soon as you walk out of the mountains and go over into the positively curved surface of the earth. But of course you can imagine a surface which preserves its negative curvature everywhere.'

'But how does it apply to a curved three-dimensional space?'

'In exactly the same way. Suppose you have objects distributed uniformly through space, I mean in such a way that the distance between two neighbouring objects is always the same, and suppose you count their number within different distances from you. If this number grows as the square of the distance, the space is flat; if the growth is slower or faster, the space possesses a positive or a negative curvature.'

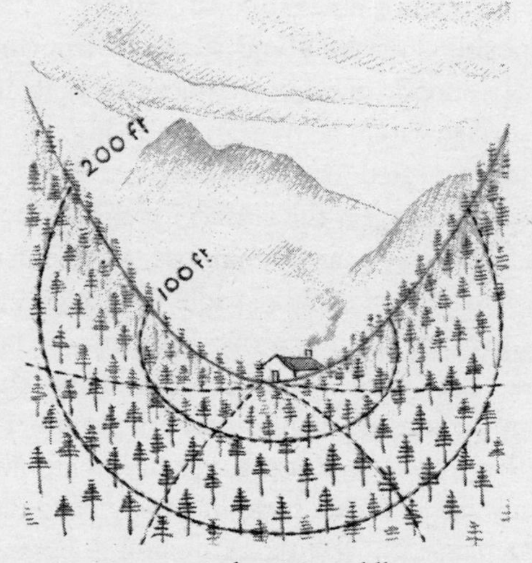

A mountain hut in a saddle pass

'Thus in the case of positive curvature the space has less volume within a given distance, and in the case of negative curvature more volume?' said Mr Tompkins with surprise.

'Just so,' smiled the professor. 'Now I see you understood me correctly. To investigate the sign of the curvature of the great universe in which we live, one just has to do such counts of the number of distant objects. The great nebulae, about which you have probably heard, are scattered uniformly through space and can be seen up to the distance of several thousand million light years; they represent very convenient objects for such investigations of the curvature of the world.'

'And so it comes out that our universe is finite and closed in itself?'

'Well,' said the professor, 'the problem is actually still unsolved. In his original papers on cosmology, Einstein stated that the universe is finite in size, closed in on itself, and unchangeable in time. Later the work of a Russian mathematician, A. A. FRIEDMANN, showed that Einstein's basic equations permit the possibility that the universe expands or contracts as it grows older. This mathematical conclusion was confirmed by an American astronomer E. HUBBLE who, using the 100-inch telescope of Mt Wilson Observatory, found that the galaxies fly apart from one another, i.e. that our universe is expanding. But there is still the problem of whether this expansion will continue indefinitely or will reach the maximum value and turn into contraction in some distant future. This question can be answered only by more detailed astronomical observations.'

While the professor was talking, very unusual changes seemed to be taking place around them: one end of the lobby became extremely small, squeezing all the furniture in it, whereas the other end was growing so large that, as it seemed to Mr Tompkins, the whole universe could find room in it. A terrible thought pierced his mind: what if a piece of space on the beach, where Miss Maud was painting, were torn away from the rest of the universe. He would never be able to see her again! When he rushed to the door he heard the professor's voice shouting behind him. 'Careful! the quantum constant is getting crazy too!' When he reached the beach it seemed to him at first very crowded. Thousands of girls were rushing in disorder in all possible directions. 'How on earth am I going to find my Maud in this crowd?' he thought. But then he noticed that they all looked exactly like the professor's daughter, and he realized that this was just the joke of the uncertainty principle. The next moment the wave of anomalously large quantum constant had passed, and Miss Maud was standing on the beach with a frightened look in her eyes.

'Oh, it is you!' she murmured with relief. 'I thought a big crowd was rushing on me. It is probably the effect of this hot sun on my head. Wait a minute until I run to the hotel and bring my sun hat.'

'Oh, no, we should not leave each other now,' protested Mr Tompkins. 'I have an impression that the velocity of light is changing too; when you return from the hotel you might find me an old man!'

'Nonsense,' said the girl, but still slipped her hand into the hand of Mr Tompkins. But half-way to the hotel another wave of uncertainty overtook them, and both Mr Tompkins and the girl spread all over the shore. At the same time a large fold of space began spreading from the hills close by, curving surrounding rocks and fishermen's houses into very funny shapes. The rays of the sun, deflected by an immense gravitational field, completely disappeared from the horizon and Mr Tompkins was plunged into complete darkness.

A century passed before a voice so dear to him brought him back to his senses.

'Oh,' the girl was saying, 'I see my father sent you to sleep by his conversation about physics. Wouldn't you like to come and have a swim with me, the water is so nice today?'

Mr Tompkins jumped from the easy chair as if on springs. 'So it was a dream after all,' he thought, as they descended towards the beach. 'Or is the dream just beginning now?'

4

The Professor's Lecture on Curved Space, Gravity and the Universe

Ladies and Gentlemen:

Today I am going to discuss the problem of curved space and its relation to the phenomena of gravitation. I have no doubt that any one of you can easily imagine a curved line or a curved surface, but at the mention of a curved, three-dimensional space your faces grow longer and you are inclined to think that it is something very unusual and almost supernatural. What is the reason for this common 'horror' for a curved space, and is this notion really more difficult than the notion of a curved surface? Many of you, if you will think a little about it, will probably say that you find it difficult to imagine a curved space because you cannot look on it 'from outside' as you look on a curved surface of a globe, or, to take another example, on the rather peculiarly curved surface of a saddle. However, those who say this convict themselves of not knowing the strict mathematical meaning of curvature, which is in fact rather different from the common use of the word. We mathematicians call a surface curved if the properties of geometrical figures drawn on it are different from those on a plane, and we measure the curvature by the deviation from the classical rules of Euclid. If you draw a triangle on a flat piece of paper the sum of its angles, as you know from elementary geometry, is equal to two right angles. You can bend this piece of paper to give to it a cylindrical, a conical, or even still more complicated shape, but the sum of the angles in the triangle drawn upon it will always remain equal to two right angles.

The geometry of the surface does not change with these deformations and, from the point of view of the 'internal' curvature, the surfaces obtained (curved in common notation) are just as flat as a plane. But you cannot fit a piece of paper, without stretching it, on to the surface of a sphere or a saddle, and, if you try to draw a triangle on a globe (i.e. a spherical triangle) the simple theorems of Euclidean geometry will not hold any more. In fact, a triangle formed, for example, by the northern halves of two meridians and a piece of the equator between them will have two right angles at its base and an arbitrary angle at the top.

On the saddle surface you will be surprised to find that, on the contrary, the sum of the angles of a triangle will always be smaller than two right angles.

Thus *to determine the curvature of a surface it is necessary to study the geometry on this surface*, whereas looking from outside will often be misleading. Just by looking you would probably place the surface of a cylinder in the same class as the surface of a ring, whereas the first is actually flat and the second is incurably curved. As soon as you get accustomed to this new strict notion of curvature you will not have any more difficulty in understanding what the physicist means in discussing whether the space in which we live is curved or not. The problem is only to find out whether the geometrical figures constructed in physical space are or are not subject to the common laws of Euclidean geometry.

Since, however, we are speaking about actual physical space we must first of all give the *physical definition of the terms used in geometry* and, in particular, state what we understand by the notion of straight lines from which our figures are to be constructed.

I suppose that all of you know that a straight line is most generally defined as the shortest distance between two points; it can be obtained either by stretching a string between two points or by an equivalent but elaborate process, of finding by trial a line between two given points along which the minimum number of measuring-sticks of given length can be placed.

In order to show that the results of such a method of finding a straight line will depend on physical conditions, let us imagine a large round platform uniformly rotating around its axis, and an experimenter (2) trying to find the shortest distance between two points on the periphery of this platform. He has a box with a large number of sticks, 5 inches each, and tries to line them up between two points so as to use the minimum total number of them. If the

The scientists were measuring something on a rotating platform*

platform were not rotating, he would place them along a line which is indicated in our figure by the dotted line. But due to the rotation of the platform his measuring-sticks will suffer a relativistic contraction, as discussed in my previous lecture, and those of them which are closer to the periphery of the platform (and therefore possess larger linear velocities) will be contracted more than those located nearer to the centre. It is thus clear that, in order to

* The name Hookham's Circus refers to Mr John Hookham, who worked as illustrator for the Cambridge University Press and, before his retirement, produced many of the drawings adorning the present volume.

get most distance covered by each stick, one should place them as close to the centre as possible. But, since both ends of the line are fixed on the periphery, it is also disadvantageous to move the sticks from the middle of the line too close to the centre.

Thus the result will be reached by a compromise between two conditions, *the shortest distance being finally represented by a curve slightly convex towards the centre.*

If, instead of using separate sticks, our experimenter will just stretch a string between the two points in question, the result will evidently be the same, because each part of the string will suffer the same relativistic contraction as the separate sticks. I want here to stress the point that this deformation of the stretched string which takes place when the platform begins to rotate has nothing to do with the usual effects of centrifugal force; in fact this deformation will not change however strongly the string is stretched, not to mention that the ordinary centrifugal force will act in the opposite direction.

If, now, the observer on the platform decides to check his results by comparing the 'straight line' he thus obtained with the ray of light, he will find that the light is really propagated along the line he has constructed. Of course, to the observers standing near the platform, the ray of light will not seem curved at all; they will interpret the results of the moving observer by the overlapping of the rotation of the platform and the rectilinear propagation of light, and will tell you that, if you make a scratch on a rotating gramophone record by moving your hand along in a straight line, the scratch on the record will also, of course, be curved.

However, as far as the observer on the rotating platform is concerned, the name of 'straight line' for the curve obtained by him is perfectly sound: it *is* the shortest distance and it *does* coincide with the ray of light in his system of reference. Suppose he now chooses three points on the periphery and connects them with straight lines, thus forming a triangle. *The sum of angles in this case will be smaller than two right angles* and he will conclude, and rightly, that the space around him is curved.

To take another example, let us suppose that two other observers on the platform (3 and 4) decide to estimate the number π by measuring the circumference of the platform and its diameter. The measuring-stick of 3 will not be affected by the rotation because its motion is always perpendicular to its length. On the other hand the stick of 4 will be always contracted and he will get for length of the periphery a value larger than for a non-rotating platform. Dividing the result of 4 by the result of 3 one will thus get *a larger value than the value of* π usually given in the text-books, which is again a result of the curvature of the space.

Not only length measurements will be affected by the rotation. A watch located on the periphery will have a large velocity and, according to the considerations of the previous lecture, will go slower than the watch standing in the centre of the platform.

If two experimenters (4 and 5) check their watches in the centre of the platform, and, after this, 5 brings his watch for some time to the periphery he will find on coming back to the centre that his watch is too slow as compared with the watch remaining all the time in the centre. He will thus conclude that in different places of the platform all physical processes go at different rates.

Suppose now our experimenters stop and think a little about the cause of the unusual results they have just obtained in their geometrical measurements. Suppose also that their platform is closed, forming a rotating room without windows, so that they could not see their motion relative to the surroundings. Could they explain all the observed results as due purely to the physical conditions on their platform without referring to its rotation relative to the 'solid ground' on which the platform is installed?

Looking for differences between the physical conditions on their platform and on the 'solid ground' by which the observed changes in the geometry could be explained, they will at once notice that there is some new force present which tends to pull all bodies from the centre of the platform towards the periphery. Naturally enough, they will ascribe the observed effects to the action of this force saying, for example, that of the two watches, the one will move slower which is further from the centre in the direction of action of this new force.

But is this force really a *new* force, not observable on the 'solid ground'? Do we not always observe that all bodies are pulled towards the centre of the earth by what is called the force of gravity? Of course, in one case we have the attraction towards the periphery of the disc, in another the attraction to the centre of the earth, but this means only a difference in the distribution of the force. It is, however, not difficult to give another example in which the 'new' forces produced by non-uniform motion of the system of reference looks exactly like the force of gravity in this lecture room.

Suppose a rocket-ship, designed for interstellar travel, floats freely somewhere in space so far from different stars that there is no force of gravity inside it. All objects inside such a rocket-ship, and the experimenter travelling in it, will thus have no weight and will float freely in the air in much the same way as Michel Ardent and his fellow-travellers to the moon in the famous story of Jules Verne.

Now the engines are being switched on, and our rocket-ship starts moving, gradually gaining velocity. What will happen inside it? It is easy to see that, as long as the ship is accelerated, all the objects in its interior will show a tendency to move towards the floor, or, to say the same thing in another way, the floor will be moving towards these objects. If, for example, our experimenter holds an apple in his hand and then lets it go, the apple will continue to move (relative to the surrounding stars) with a constant velocity—the velocity with which the rocket-ship was moving at the moment when the apple was released. But the rocket-ship itself is accelerated; consequently the floor of the cabin, moving all the time faster and faster, will finally overtake the apple and hit it; from this moment on the apple will remain permanently in contact with the floor, being pressed to it by steady acceleration.

For the experimenter inside, however, this will look as if the apple 'falls down' with a certain acceleration, and after hitting the floor remains pressed to it by its own weight. Dropping different objects, he will notice furthermore that all of them fall with exactly equal accelerations (if he neglects the friction of the air) and will remember that this is exactly the rule of the free fall discovered by GALILEO GALILEI. *In fact he will not be able to notice the slightest difference between the phenomena in his accelerated cabin and the ordinary phenomena of gravity.* He can use the clock with the pendulum, put books on a shelf without any danger of their flying away, and hang on a nail the portrait of Albert Einstein, who first indicated the equivalence of acceleration of the system of reference and the field of gravity, and developed, on this basis, the so-called general theory of relativity.

But here, just as in the first example of a rotating platform, we shall notice phenomena unknown to Galileo and Newton in their study of gravity. The light ray sent across the cabin will get curved and will illuminate a screen hanging on the opposite wall at different places, depending on the acceleration of the rocket-ship. By an outside observer, this will be interpreted, of course, as due to the overlapping of a uniform rectilinear motion of light and the accelerated motion of the observational cabin. The geometry will also go wrong; the sum of angles of a triangle formed by three light rays will be larger than two right angles, and the ratio of the periphery of a circle to its diameter will be larger than the number π. We have considered here two of the simplest examples of accelerated systems, but the equivalence stated above will hold for any given motion of a rigid or a deformable system of reference.

We come now to the question of greatest importance. We have just seen that in an accelerated system of reference a number of phenomena could be observed that were unknown for the ordinary

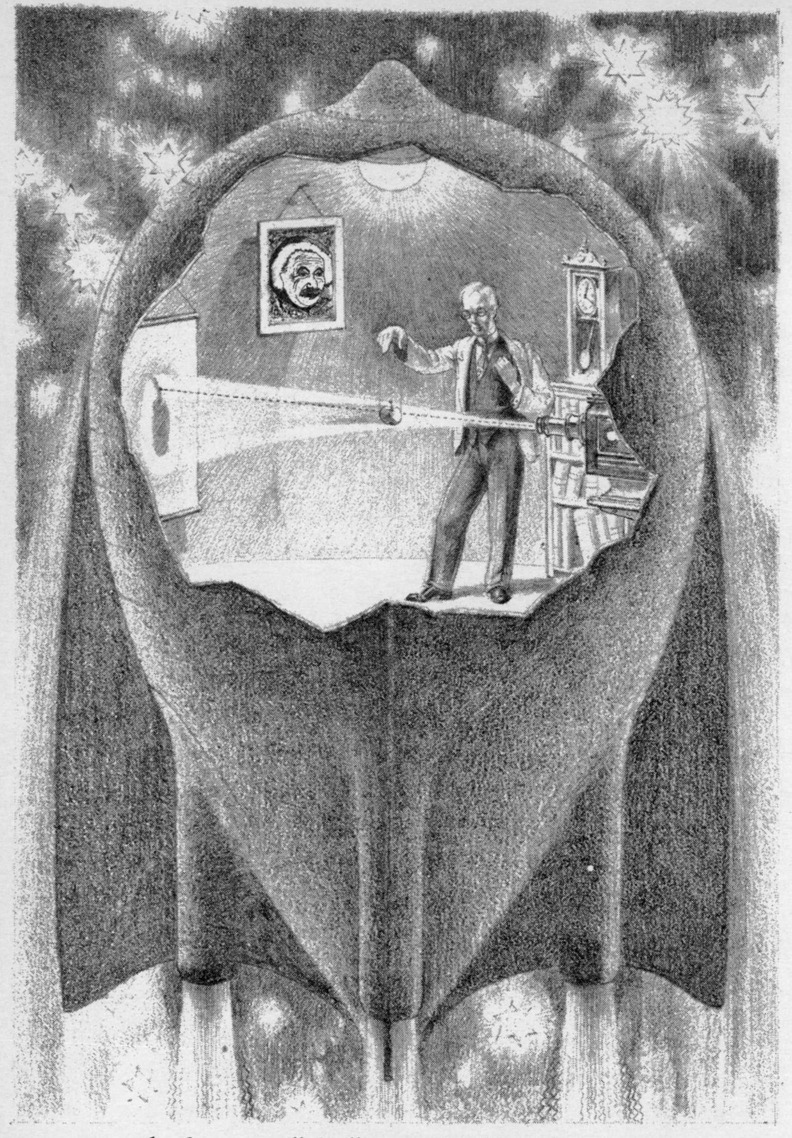

The floor...will finally overtake the apple and hit it

field of gravitation. Do these new phenomena, such as the curving of a light ray or slowing down of a clock, also exist in gravitational fields produced by ponderable masses? Or, in other words, are the effects of acceleration and the effects of gravity not only very similar, but identical?

It is clear, of course, that although from the heuristic point of view it is very tempting to accept complete identity of these two kinds of effects, the final answer can be given only by direct experiments. And, to the great satisfaction of our human mind, which demands simplicity and internal consistency of the laws of the universe, experiments do prove the existence of these new phenomena also in the ordinary field of gravity. Of course, the effects predicted by the hypothesis of the equivalence of accelerative and gravitational fields are very small: that is why they have been discovered only after scientists started looking specially for them.

Using the example of accelerated systems discussed above, we can easily estimate the order of magnitude of the two most important relativistic gravitational phenomena: the change of the clock rate and the curvature of a light ray.

Let us first take the example of the rotating platform. It is known from elementary mechanics that the centrifugal force acting on a particle of mass unity located at the distance r from the centre is given by the formula

$$F = r\omega^2, \tag{1}$$

where ω is the constant angular velocity of rotation of our platform. The total work done by this force during the motion of the particle from the centre to the periphery is then

$$W = \tfrac{1}{2}R^2\omega^2, \tag{2}$$

where R is the radius of the platform.

According to the above-stated equivalence principle, we have to identify F with the force of gravity on the platform, and W

with the difference of gravitational potential between the centre and the periphery.

Now, we must remember that, as we have seen in the previous lecture, the slowing down of the clock moving with the velocity v is given by the factor

$$\sqrt{1-\left(\frac{v}{c}\right)^2} = 1-\frac{1}{2}\left(\frac{v}{c}\right)^2+\ldots. \tag{3}$$

If v is small as compared with c we can neglect other terms. According to the definition of the angular velocity we have $v = R\omega$ and the 'slowing-down factor' becomes

$$1-\frac{1}{2}\left(\frac{R\omega}{c}\right)^2 = 1-\frac{W}{c^2}, \tag{4}$$

giving the change of rate of the clock in terms of the difference of gravitational potentials at the places of their location.

If we place one clock at the basement and another on the top of the Eiffel tower (1000 feet high) the difference of potential between them will be so small that the clock at the basement will go slower only by a factor 0·999,999,999,999,97.

On the other hand, the difference of gravitational potential between the surface of the earth and the surface of the sun is much larger, giving the slowing down by a factor 0·999,999,5, which can be noticed by very exact measurements. Of course, nobody was going to place an ordinary clock on the surface of the sun and watch it go! The physicists have much better means. By means of the spectroscope we can observe the periods of vibration of different atoms on the surface of the sun and compare them with the periods of the atoms of the same elements put into the flame of a Bunsen-burner in the laboratory. The vibrations of atoms on the surface of the sun should be slowed down by the factor given by the formula (4) and the light emitted by them should be somewhat more reddish than in the case of terrestrial sources. This 'red-shift' was actually observed in the spectra of the sun and several other stars, for which the spectra could be exactly measured, and the result agrees with the value given by our theoretical formula.

Thus the existence of the red-shift proved that the processes on the sun really take place somewhat more slowly owing to higher gravitational potential on its surface.

In order to get a measure for the curvature of a light ray in the field of gravity it is more convenient to use the example of the rocket-ship as given on page 36. If l is the distance across the cabin, the time t taken by light to cross it is given by

$$t = \frac{l}{c}. \tag{5}$$

During this time the ship, moving with the acceleration g, will cover the distance L given by the following formula of elementary mechanics:

$$L = \tfrac{1}{2}gt^2 = \tfrac{1}{2}g\frac{l^2}{c^2}. \tag{6}$$

Thus the angle representing the change of the direction of the light ray is of the order of magnitude

$$\phi = \frac{L}{l} = \frac{1}{2}\frac{gl}{c^2} \text{ radians}, \tag{7}$$

and is larger, the larger the distance l which the light has travelled in the gravitational field. Here the acceleration g of the rocket-ship has, of course, to be interpreted as the acceleration of gravity. If I send a beam of light across this lecture room, I can take roughly $l = 1000$ cm. The acceleration of gravity g on the surface of the earth is 981 cm/sec^2 and with $c = 3 \cdot 10^{10}$ cm/sec we get

$$\phi = \frac{100 \times 981}{2 \times (3 \cdot 10^{10})^2} = 5 \cdot 10^{-16} \text{ radians} = 10^{-10} \text{ sec of arc}. \tag{8}$$

Thus you can see that the curvature of light can definitely not be observed under such conditions. However, near the surface of the sun g is 27,000 and the total path travelled in the gravitational field of the sun is very large. The exact calculations show that the value for the deviation of a light ray passing near the solar surface should be 1·75″, and this is just exactly the value observed by astronomers for the displacement of the apparent position of stars seen near the solar limb during a total eclipse. You see that here, too, the observations have shown a complete identity of the effects of acceleration and those of gravitation.

Now we can return again to our problem about the curvature of space. You remember that, using the most rational definition of a straight line, we came to the conclusion that the geometry obtained in ununiformly moving systems of reference is different from that of Euclid and that such spaces should be considered as curved spaces. Since any gravitational field is equivalent to some acceleration of the system of reference, this means also that any space in which the gravitational field is present is a curved space. Or, going still a step farther, that *a gravitational field is just a physical manifestation of the curvature of space*. Thus the curvature of space at each point should be determined by the distribution of masses, and near heavy bodies the curvature of space should reach its maximum value. I cannot enter into a rather complicated mathematical system describing the properties of curved space and their dependence on the distribution of masses. I should mention only that this curvature is in general determined not by one, but by ten different numbers which are usually known as the components gravitational potential $g_{\mu\nu}$ and represent a generalization of the gravitational potential of classical physics which I have previously called W. Correspondingly, the curvature at each point is described by ten different radii of curvature usually denoted by $R_{\mu\nu}$. Those radii of curvature are connected with distribution of masses by the fundamental equation of Einstein:

$$R_{\mu\nu} - \tfrac{1}{2} g_{\mu\nu} R = -\kappa T_{\mu\nu}, \qquad (9)$$

where $T_{\mu\nu}$ depends on densities, velocities and other properties of the gravitational field produced by ponderable masses.

Coming to the end of this lecture, I should like, however, to indicate one of the most interesting consequences of equation (9). If we consider a space uniformly filled with masses, as, for example, our space is filled with stars and stellar systems, we shall come to the conclusion that, apart from occasionally large curvatures near separate stars, the space should possess *a regular tendency to curve uniformly on large distances*. Mathematically there are several different solutions, some of them corresponding to the *space finally closing on itself and thus possessing a finite volume*, the others representing *the infinite space analogous to a saddle surface* which I mentioned at the beginning of this lecture. The second important consequence of equation (9) is that such curved spaces should be in a state of steady expansion or contraction, which physically means that the particles filling the space should be flying away from each other, or, on the contrary, approaching each other. Further, it can be shown that for the closed spaces with finite volume the expansion and contraction periodically follow each other—these are so-called pulsating worlds. On the other hand, infinite 'saddle-like' spaces are permanently in a state of contraction or of expansion.

The question which of all these different mathematical possibilities corresponds to the space in which we are living should be answered not by physics but by astronomy and I am not going to discuss it here. I will mention only that so far astronomical evidence has definitely shown that our space is expanding, although the question whether this expansion will ever turn into a contraction, and whether the space is finite or infinite in size is not yet definitely settled.

5

The Pulsating Universe

After dinner on their first evening in the Beach Hotel with the old professor talking about cosmology, and his daughter chatting about art, Mr Tompkins finally got to his room, collapsed on to the bed, and pulled the blanket over his head. Botticelli and Bondi, Salvador Dali and Fred Hoyle, Lemaître and La Fontaine got all mixed up in his tired brain, and finally he fell into a deep sleep....

Sometime in the middle of the night he woke up with a strange feeling that instead of lying on a comfortable spring mattress he was lying on something hard. He opened his eyes and found himself prostrated on what he first thought to be a big rock on the seashore. Later he discovered that it was actually a very big rock, about 30 feet in diameter, suspended in space without any visible support. The rock was covered with some green moss, and in a few places little bushes were growing from cracks in the stone. The space around the rock was illuminated by some glimmering light and was very dusty. In fact, there was more dust in the air than he had ever seen, even in the films representing dust storms in the middle west. He tied his handkerchief round his nose and felt, after this, considerably relieved. But there were more dangerous things than the dust in the surrounding space. Very often stones of the size of his head and larger were swirling through the space near his rock, occasionally hitting it with a strange dull sound of impact. He noticed also one or two rocks of approximately the same size as his own, floating through space at some distance away. All this time, inspecting his surroundings, he was clinging hard to some protruding edges of his rock in constant fear of falling off and being lost in the dusty depths below. Soon, however, he became bolder, and made an attempt to crawl to the edge of his rock and to see whether there was really nothing underneath, supporting it. As he was crawling in this way, he noticed, to his great surprise, that he did not fall off, but that his weight was constantly pressing him to the surface of the rock, although he had covered already more than a quarter of its circumference. Looking from behind a ridge of loose stones on the spot just underneath the place where he originally found himself, he discovered nothing to support the rock in space. To his great surprise, however, the glimmering light revealed the tall figure of his friend the old professor standing apparently with his head down and making some notes in his pocket-book.

Now Mr Tompkins began slowly to understand. He remembered that he was taught in his schooldays that the earth is a big round rock moving freely in space around the sun. He also remembered the picture of two antipodes standing on the opposite sides of the earth. Yes, his rock was just a very small stellar body attracting everything to its surface, and he and the old professor were the only population of this little planet. This consoled him a little; there was at least no danger of falling off!

'Good morning,' said Mr Tompkins, to divert the old man's attention from his calculations.

The professor raised his eyes from his note-book. 'There are no mornings here,' he said, 'there is no sun and not a single luminous star in this universe. It is lucky that the bodies here show some chemical process on their surface, otherwise I should not be able to observe the expansion of this space', and he returned again to his note-book.

Mr Tompkins felt quite unhappy; to meet the only living person in the whole universe, and to find him so unsociable! Unexpectedly, one of the little meteorites came to his help; with a crashing sound the stone hit the book in the hands of the professor and threw it, travelling fast through space, away from their little planet. 'Now you will never see it again,' said Mr Tompkins, as the book got smaller and smaller, flying through space.

'On the contrary,' replied the professor. 'You see, the space in which we now are is not infinite in its extension. Oh yes, yes, I know that you have been taught in school that space is infinite, and that two parallel lines never meet. This, however, is not true either for the space in which the rest of humanity lives, or for the space in which we are now. The first one is of course very large indeed; the scientists estimated its present dimensions to be about 10,000,000,000,000,000,000,000 miles, which, for an ordinary

There are no mornings here

mind, is fairly infinite. If I had lost my book there, it would take an incredibly long time to come back. Here, however, the situation is rather different. Just before the note-book was torn out of my hands, I had figured out that this space is only about five miles in diameter, though it is rapidly expanding. I expect the book back in not more than half an hour.'

'But,' ventured Mr Tompkins, 'do you mean that your book is going to behave like the boomerang of an Australian native, and, by moving along a curved trajectory, fall down at your feet?'

'Nothing of the sort,' answered the professor. 'If you want to understand what really happens, think about an ancient Greek who did not know that the earth was a sphere. Suppose he has given somebody instructions to go always straight northwards. Imagine his astonishment when his runner finally returns to him from the south. Our ancient Greek did not have a notion about travelling round the world (round the earth, I mean in this case), and he would be sure that his runner had lost his way and had taken a curved route which brought him back. In reality his man was going all the time along the straightest line one can draw on the surface of the earth, but he travelled round the world and thus came back from the opposite direction. The same thing is going to happen to my book, unless it is hit on its way by some other stone and thus deflected from the straight track. Here, take these binoculars, and see if you can still see it.'

Mr Tompkins put the binoculars to his eyes, and, through the dust which somewhat obscured the whole picture, he managed to see the professor's note-book travelling through space far far away. He was somewhat surprised by the pink colouring of all the objects, including the book, at that distance.

'But,' he exclaimed after a while, 'your book is returning, I see it growing larger.'

'No,' said the professor, 'it is still going away. The fact that you see it growing in size, as if it were coming back, is due to a peculiar focusing effect of the closed spherical space on the rays of light. Let us return to our ancient Greek. If the rays of light could be kept going all the time along the curved surface of the earth, let us say by refraction of the atmosphere, he would be able, using powerful binoculars, to see his runner all the time during the journey. If you look on the globe, you will see that the straightest lines on its surface, the meridians, first diverge from one pole, but, after passing the equator, begin to converge towards the opposite pole. If the rays of light travelled along the meridians, you, located, for example, at one pole, would see the person going away from you growing smaller and smaller only until he crossed the equator. After this point you would see him growing larger and it would seem to you that he was returning, going, however, backwards. After he had reached the opposite pole, you would see him as large as if he were standing right by your side. You would not be able, however, to touch him, just as you cannot touch the image in a spherical mirror. On this basis of two-dimensional analogy, you can imagine what happens to the light rays in the strangely curved three-dimensional space. Here, I think the image of the book is quite close now.' In fact, dropping the binoculars, Mr Tompkins could see that the book was only a few yards away. It looked, however, very strange indeed! The contours were not sharp, but rather washed out, the formulae written by the professor on its pages could be hardly recognized, and the whole book looked like a photograph taken out of focus and underdeveloped.

'You see now,' said the professor, 'that this is only the image of the book, badly distorted by light travelling across one half of the universe. If you want to be quite sure of it, just notice how the stones behind the book can be seen through its pages.'

Mr Tompkins tried to reach the book, but his hand passed through the image without any resistance.

'The book itself,' said the professor, 'is now very close to the opposite pole of the universe, and what you see here are just two images of it. The second image is just behind you and when both images coincide, the real book will be exactly at the opposite pole.' Mr Tompkins didn't hear; he was too deeply absorbed in his thoughts, trying to remember how the images of objects are formed in elementary optics by concave mirrors and lenses. When he finally gave it up, the two images were again receding in opposite directions.

'But what makes the space curved and produce all these funny effects?' he asked the professor.

'The presence of ponderable matter,' was the answer. 'When

Newton discovered the law of gravity, he thought that gravity was just an ordinary force, the same type of force as, for example, is produced by an elastic string stretched between two bodies. There always remains, however, the mysterious fact that all bodies, independent of their weight and size, have the same acceleration and move the same way under the action of gravity, provided you eliminate the friction of air and that sort of thing, of course. It was Einstein who first made it clear that the primary action of ponderable matter is to produce the curvature of space and that the trajectories of all bodies moving in the field of gravity are curved just because space itself is curved. But I think it is too hard for you to understand, without knowing sufficient mathematics.'

'It is,' said Mr Tompkins. 'But tell me, if there were no matter, would we have the kind of geometry I was taught at school, and would parallel lines never meet?'

'They would not,' answered the professor, 'but neither would there be any material creature to check it.'

'Well, perhaps Euclid never existed, and therefore could construct the geometry of absolutely empty space?'

But the professor apparently did not like to enter into this metaphysical discussion.

In the meantime the image of the book went off again far away in the original direction, and started coming back for the second time. Now it was still more damaged than before, and could hardly be recognized at all, which, according to the professor, was due to the fact that the light rays had travelled this time round the whole universe.

'If you turn your head once more,' he said to Mr Tompkins, 'you will see my book finally coming back after completing its journey round the world.' He stretched his hand, caught the book, and pushed it into his pocket. 'You see,' he said, 'there is so much dust and stone in this universe that it makes it almost impossible to see round the world. These shapeless shadows which you might notice around us are most probably the images of ourselves, and surrounding objects. They are, however, so much distorted by dust and irregularities of the curvature of space that I cannot even tell which is which.'

'Does the same effect occur in the big universe in which we used to live before?' asked Mr Tompkins.

'Oh yes,' was the answer, 'but that universe is so big that it takes the light milliards of years to go round. You could have seen the hair cut on the back of your head without any mirror, but only milliards of years after you had been to the barber. Besides, most probably the interstellar dust would completely obscure the picture. By the way, one English astronomer even supposed once, mostly as a joke, that some of the stars which can be seen in the sky at present are only the images of stars which existed long ago.'

Tired of the efforts to understand all these explanations, Mr Tompkins looked around and noticed, to his great surprise, that the picture of the sky had considerably changed. There seemed to be less dust about, and he took off the handkerchief which was still tied round his face. The small stones were passing much less frequently and hitting the surface of their rock with much less energy. Finally, a few big rocks like their own, which he had noticed in the very beginning, had gone much farther away and could hardly be seen at this distance.

'Well, life is certainly becoming more comfortable,' thought Mr Tompkins. 'I was always so scared that one of those travelling stones would hit me. Can you explain the change in our surroundings?' he said, turning to the professor.

'Very easily; our little universe is rapidly expanding and since we have been here its dimensions have increased from *five to about a hundred miles*. As soon as I found myself here, I noticed this expansion from the reddening of the distant objects.'

'Well, I also see that everything is getting pink, at great distances,' said Mr Tompkins, 'but why does it signify expansion?'

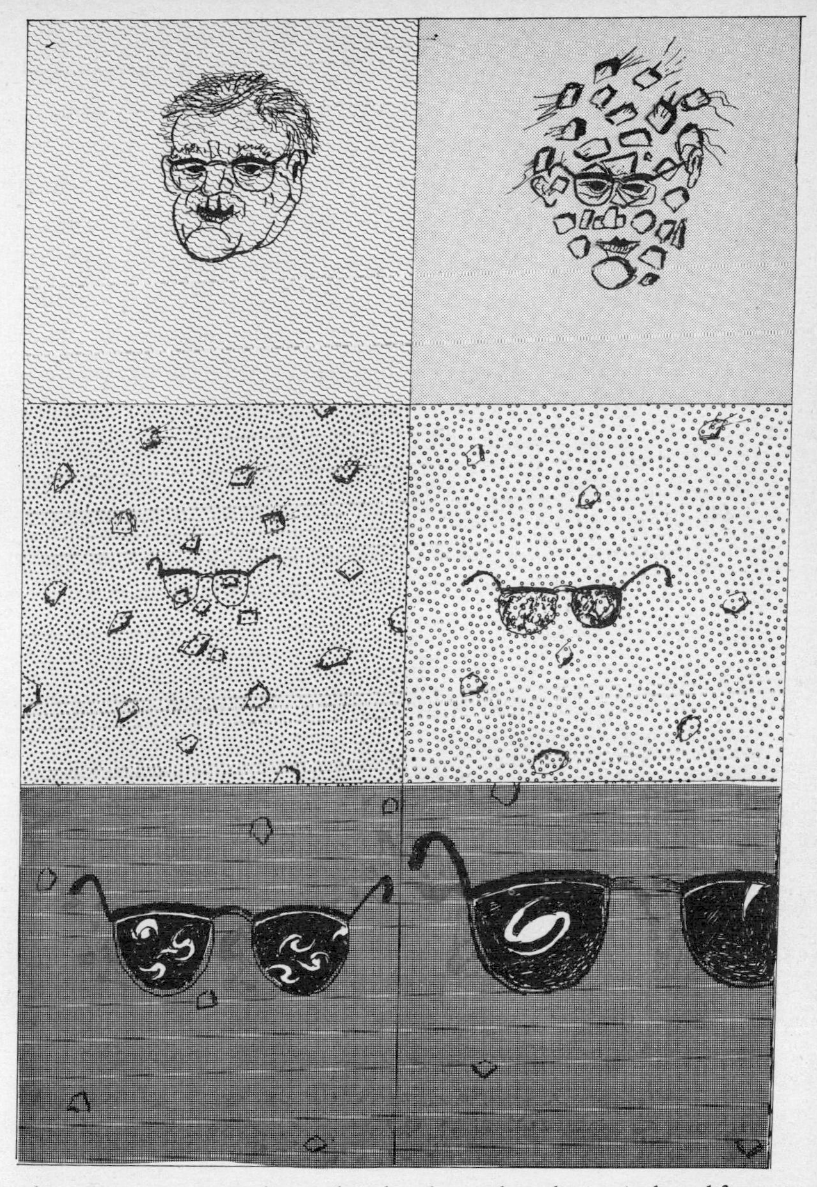

The universe was expanding and cooling beyond any limit. (Adapted from a cartoon in *The Sydney Daily Telegraph*, 16 January 1960)

'Have you ever noticed,' said the professor, 'that the whistle of an approaching train sounds very high, but after the train passes you, the tone is considerably lower? This is the so-called Doppler Effect: the dependence of the pitch on the velocity of the source. When the whole space is expanding, every object located in it moves away with a velocity proportional to its distance from the observer. Therefore the light emitted by such objects is getting redder, which in optics corresponds to a lower pitch. The more distant the object is, the faster it moves and the redder it seems to us. In our good old universe, which is also expanding, this reddening, or the red-shift as we call it, permits astronomers to estimate the distances of the very remote clouds of stars. For example, one of the nearest clouds, the so-called Andromeda nebula, shows 0·05 % of reddening, which corresponds to the distance which can be covered by light in eight hundred thousand years. But there are also nebulae just on the limit of present telescopic power, which show a reddening of about 15 % corresponding to distances of several hundred millions of light years. Presumably, these nebulae are located almost on the half-way point of the equator of the big universe, and the total volume of space which is known to terrestrial astronomers represents a considerable part of the total volume of that universe. The present rate of expansion is about 0·000,000,01 % per year, so that each second the radius of the universe increases by *ten million* miles. Our little universe grows comparatively much faster, gaining in its dimensions about 1 % per minute.'

'Will this expansion never stop?' asked Mr Tompkins.

'Of course it will,' said the professor. 'And then the contraction will start. Each universe pulsates between a very small and a very large radius. For the big universe the period is rather large, something like several thousand million years, but our little one has a period of only about two hours. I think we are now observing the state of largest expansion. Do you notice how cold it is?'

In fact, the thermal radiation filling up the universe, and now distributed over a very large volume, was giving only very little heat to their little planet, and the temperature was at about freezing-point.

'It is lucky for us,' said the professor, 'that there was originally enough radiation to give some heat even at this stage of expansion. Otherwise it might become so cold that the air around our rock would condense into liquid and we would freeze to death. But the contraction has already begun, and it will soon be warm again.'

Looking at the sky, Mr Tompkins noticed that all distant objects changed their colour from pink to violet which, according to the professor, was due to the fact that all the stellar bodies had started moving towards them. He also remembered the analogy given by the professor of the high pitch of the whistle of an approaching train, and shuddered from fear.

'If everything is contracting now, shouldn't we expect that soon all the big rocks filling the universe will come together and that we shall be crushed between them?' he asked the professor anxiously.

'Exactly so,' answered the professor calmly, 'but I think that even before this the temperature will rise so high that we shall both be dissociated into separate atoms. This is a miniature picture of the end of the big universe—everything will be mixed up into a uniform hot gas sphere, and only with a new expansion will new life begin again.'

'Oh my!' muttered Mr Tompkins—'In the big universe we have, as you mentioned, milliards of years before the end, but here it is going too fast for me! I feel hot already, even in my pyjamas.'

'Better not take them off,' said the professor, 'it will not help. Just lie down and observe as long as you can.'

Mr Tompkins did not answer; the hot air was unbearable. The dust, which became very dense now, was accumulating around him, and he felt as if he were being rolled up in a soft warm blanket. He made a motion to free himself, and his hand came out into cool air.

'Did I make a hole in that inhospitable universe?' was his first thought. He wanted to ask the professor about it, but could not

find him anywhere. Instead, in the dim light of the morning, he recognized the contours of the familiar bedroom furniture. He was lying in his bed tightly rolled up in a woollen blanket, and had just managed to free one hand from it.

'New life begins with expansion,' he thought, remembering the words of the old professor. 'Thank God we are still expanding!' And he went to take his morning bath.

6

Cosmic Opera

When, that morning at breakfast, Mr Tompkins told the professor about his dream the previous night, the old man listened rather sceptically.

'The collapse of the universe,' said he, 'would of course be a very dramatic ending, but I think that the velocities of mutual recession of galaxies are so high that present expansion will never turn into a collapse, and that the universe will continue to expand beyond any limit with the distribution of galaxies in space becoming more and more diluted. When all the stars forming the galaxies burn out because of the exhaustion of nuclear fuel, the universe will become a collection of cold and dark celestial aggregations dispersing into infinity.'

'There are, however, some astronomers who think otherwise. They suggest the so-called steady state cosmology, according to which the universe remains unchanging in time: it has existed in about the same state as we see it today from infinity in the past, and will continue so to exist to infinity in the future. Of course it is in accordance with the good old principle of the British empire to preserve the status quo in the world, but I am not inclined to believe that this steady state theory is true. By the way, one of the originators of this new theory, a professor of theoretical astronomy at Cambridge University, wrote an opera on the subject which will have its premiere in Covent Garden next week. Why don't you reserve tickets for Maud and yourself and go to hear it? It may be quite amusing.'

A few days after returning from the beach, which like most channel beaches becomes cool and rainy, Mr Tompkins and Maud were resting comfortably in the red velvet chairs of the opera house, waiting for the curtain to rise. The prelude began *precipite-*

volissimevolmente, and the orchestra leader had to change the collar of his dress suit twice before it was over. When finally the curtain was jerked up, everybody in the audience had to shade his eyes with the palms of his hands, so brilliant was the illumination of the stage. The intense beams of light emanating from the stage

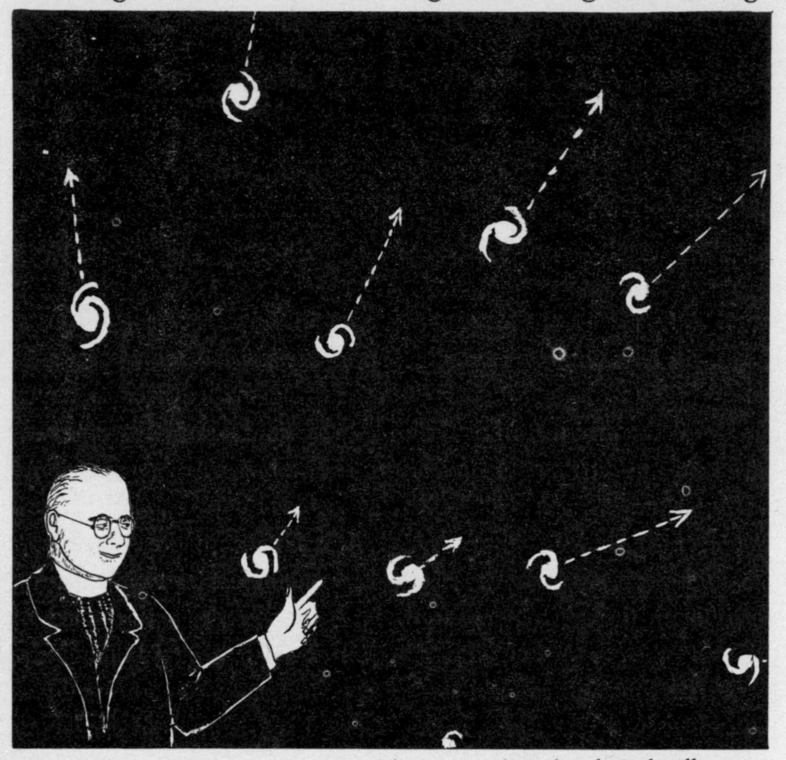

Mr Tompkins saw a man in a black cassock and a clerical collar

soon filled the entire hall, and the ground floor as well as the balcony became one brilliant ocean of light. Gradually the general brilliance faded out, and Mr Tompkins found himself apparently floating in darkened space, illuminated by a multitude of rapidly rotating flaming torches resembling the firewheels used at night festivals. The music of the invisible orchestra now began to sound like organ music and Mr Tompkins saw near him a man in a black cassock and a clerical collar. According to the libretto, it was Abbé Georges Lemaître from Belgium who was the first to propose the theory of the expanding universe, which one often calls the 'big bang' theory.

Mr Tompkins still remembers the first stanzas of his aria:

O, Atome prreemorrdiale!
All-containeeng Atome!
Deessolved eento frragments exceedeengly small.
  Galaxies forrmeeng,
  Each wiz prrimal enerrgy!
O, rradioactif Atome!
O, all-containeeng Atome!
O, Univairrsale Atome—
  Worrk of z' Lorrd!

Z' long evolution
Tells of mighty firreworrks
Zat ended een ashes and smouldairreeng weesps.
We stand on z' ceendairres
Fadeeng suns confrronteeng us,
Attempteeng to rremembairre
Z' splendeur of z' origine.
O, Univairrsale Atome—
Worrk of Z' Lorrd!

After Father Lemaître finished his aria, there appeared a tall fellow who (according to the libretto again) was a Russian physicist, George Gamow, who had been taking his vacation in the United States for the last three decades. This is what he sang:

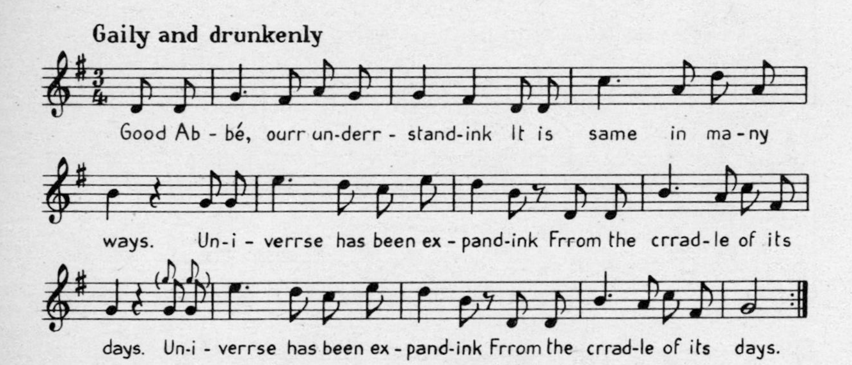

Good Abbé, ourr underrstandink
It is same in many ways.
Univerrse has been expandink
Frrom the crradle of its days.
Univerrse has been expandink
Frrom the crradle of its days.

You have told it gains in motion.
I rregrret to disagrree,
And we differr in ourr notion
As to how it came to be.
And we differr in ourr notion
As to how it came to be.

It was neutrron fluid—neverr
Prrimal Atom, as you told.
It is infinite, as everr
It was infinite of old.
It is infinite, as everr
It was infinite of old.

On a limitless pavilion
In collapse, gas met its fate,
Yearrs ago (some thousand million)
Having come to densest state.
Yearrs ago (some thousand million)
Having come to densest state.

All the Space was then rresplendent
At that crrucial point in time.
Light to matterr was trranscendent
Much as meterr is, to rrhyme.
Light to matterr was trranscendent
Much as meterr is, to rrhyme.

Forr each ton of rradiation
Then of matterr was an ounce,
Till the impulse t'warrd inflation
In that grreat prrimeval bounce.
Till the impulse t'warrd inflation
In that grreat prrimeval bounce.

Light by then was slowly palink.
Hundrred million yearrs go by...
Matterr, over light prrevailink,
Is in plentiful supply.
Matterr, over light prrevailink,
Is in plentiful supply.

Matterr then began condensink
(Such are Jeans' hypotheses).
Giant, gaseous clouds dispensink
Known as prrotogalaxies.
Giant, gaseous clouds dispensink
Known as prrotogalaxies.

Prrotogalaxies were shatterred,
Flying outward thrrough the night.

Starrs werre forrmed frrom them, and scatterred
And the Space was filled with light.
Starrs werre forrmed frrom them, and scatterred
And the Space was filled with light.

Galaxies arre everr spinnink,
Starrs will burrn to final sparrk,
Till ourr universe is thinnink
And is lifeless, cold and darrk.
Till ourr universe is thinnink
And is lifeless, cold and darrk.

The third aria which Mr Tompkins remembers was delivered by the author of the opera himself, who suddenly materialized from nothing in the space between the brightly shining galaxies. He was pulling a newborn galaxy from his pocket and singing:

Refrain

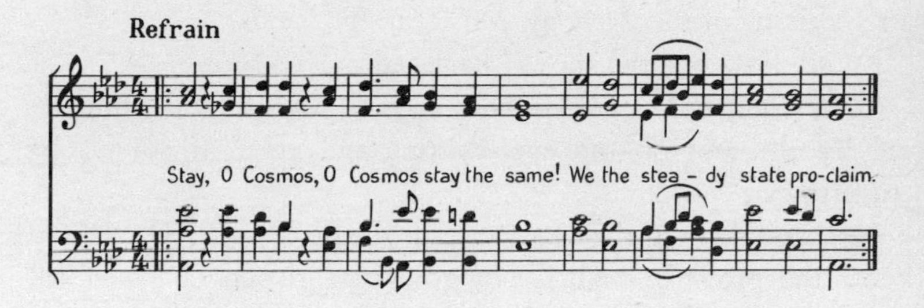

The universe, by Heaven's decree,
Was never formed in time gone by,
But is, has been, shall ever be—
For so say Bondi, Gold and I.
Stay, O Cosmos, O Cosmos, stay the same!
We the Steady State proclaim!

The aging galaxies disperse,
Burn out, and exit from the scene.
But all the while, the universe
Is, was, shall ever be, has been.
Stay, O Cosmos, O Cosmos, stay the same!
We the Steady State proclaim!

And still new galaxies condense
From nothing, as they did before.
(Lemaître and Gamow, no offence!)
All was, will be for evermore.
Stay, O Cosmos, O Cosmos, stay the same!
We the Steady State proclaim!

But in spite of these inspiring words all the galaxies in the surrounding space were gradually fading out, and finally the velvet curtain was lowered and the candelabra in the large opera hall took their place.

'Oh, Cyril,' he heard Maud say, 'I know you are apt to fall asleep in any place at any time, but you shouldn't in Covent Garden! You slept through the entire performance!'

When Mr Tompkins brought Maud back to her father's house the professor was sitting in his comfortable chair with the newly arrived issue of the *Monthly Notices* in his hands.

'Well, how was the show?' he asked.

'Oh, wonderful!' said Mr Tompkins, 'I was especially impressed by the aria on the ever-existing universe. It sounds so reassuring.'

'Be careful about this theory,' said the professor. 'Don't you know the proverb: "All is not gold that glitters"? I am just reading an article by another Cambridge man, MARTIN RYLE, who built a giant radio-telescope which can locate galaxies at distances several times greater than the range of the Mount Palomar 200-inch optical telescope. His observations show that these very distant galaxies are located much closer to each other than are those in our neighbourhood.'

'Do you mean,' asked Mr Tompkins, 'that our region of the universe has a rather rare population of galaxies, and that this population density increases when we go further and further away?'

'Not at all,' said the professor, 'you must remember that, due to the finite velocity of light, when you look far out into space you look also far back into time. For example, since light takes eight minutes to come here from the Sun, a flare on the Sun's surface is observed by terrestrial astronomers with an eight-minute delay. The photographs of our nearest space neighbour, a spiral galaxy in the constellation of Andromeda—which you must have seen in books on astronomy and which is located about one million light-years away—show how it actually looked one million years ago. Thus, what Ryle sees, or should I rather say hears, through his radio-telescope, corresponds to the situation which existed in that distant part of the universe many thousand millions of years ago. If the universe were really in a steady state, the picture should be unchanged in time, and very distant galaxies as observed from here now should be seen distributed in space neither more densely nor rarely than the galaxies at shorter distances. Thus Ryle's observations showing that distant galaxies seem to be more closely packed together in space is equivalent to the statement that the galaxies everywhere were packed more closely together in the distant past of thousands of millions of years ago. This contradicts the steady state theory, and supports the original view that the galaxies are dispersing and that their population density is going down. But of course we must be careful and wait for further confirmation of Ryle's results.'

'By the way,' continued the professor, extracting a folded piece of paper from his pocket, 'here is a verse which one of my poetically inclined colleagues wrote recently on this subject.' And he read:

'Your years of toil,'
Said Ryle to Hoyle,
  'Are wasted years, believe me.
The steady state
Is out of date.
  Unless my eyes deceive me,

My telescope
Has dashed your hope;
  Your tenets are refuted.
Let me be terse:
Our universe
  Grows daily more diluted!'

Said Hoyle, 'You quote
Lemaître, I note,
  And Gamow. Well, forget them!
That errant gang
And their Big Bang—
  Why aid them and abet them?

You see, my friend,
It has no end
And there was no beginning,
As Bondi, Gold,
And I will hold
Until our hair is thinning!'

'Not so!' cried Ryle
With rising bile
And straining at the tether;
'*Far galaxies*
Are, as one sees,
More tightly packed together!'

'You make me boil!'
Exploded Hoyle,
His statement rearranging;
'*New matter's born*
Each night and morn.
The picture is unchanging!'

'Come off it, Hoyle!
I aim to foil
You yet' (The fun commences)
'And in a while,'
Continued Ryle,
'I'll bring you to your senses!' *

'Well,' said Mr Tompkins, 'it will be exciting to see what will be the outcome of this dispute,' and giving Maud a kiss on the cheek he wished them both goodnight.

*A fortnight before the publication date of the first printing of this book there appeared an article by F. Hoyle entitled: "Recent Developments in Cosmology" (*Nature*, Oct. 9, 1965, p. 111). Hoyle writes: "Ryle and his associates have counted radio sources . . . The indication of that radio count is that the Universe was more dense in the past than it is today." The author has decided, however, not to change the lines of the arias of "Cosmic Opera" since, once written, operas become classic. In fact, even today Desdemona sings a beautiful aria before she dies, after being strangled by Othello.

7
Quantum Billiards

One day Mr Tompkins was going home, feeling very tired after the long day's work in the bank, which was doing a land office business. He was passing a pub and decided to drop in for a glass of ale. One glass followed the other, and soon Mr Tompkins began to feel rather dizzy. In the back of the pub was a billiard room filled with men in shirt sleeves playing billiards on the central table. He vaguely remembered being here before, when one of his fellow clerks took him along to teach him billiards. He approached the table and started to watch the game. Something very queer about it! A player put a ball on the table and hit it with the cue. Watching the rolling ball, Mr Tompkins noticed to his great surprise that the ball began to 'spread out'. This was the only expression he could find for the strange behaviour of the ball which, moving across the green field, seemed to become more and more washed out, losing its sharp contours. It looked as if not one ball was rolling across the table but a great number of balls, all partially penetrating into each other. Mr Tompkins had often observed analogous phenomena before, but today he had not taken a single drop of whisky and he could not understand why it was happening now. 'Well,' he thought, 'let us see how this gruel of a ball is going to hit another one.'

The player who hit the ball was evidently an expert and the rolling ball hit another one head-on just as it was meant to. There was a loud sound of impact and both the resting and the incident balls (Mr Tompkins could not positively say which was which) rushed 'in all different directions'. Yes, it was very strange; there were no longer two balls looking only somewhat gruelly, but instead it seemed that innumerable balls, all of them *very* vague and gruelly, were rushing about within an angle of 180° round the direction of the original impact. It resembled rather a peculiar wave spreading from the point of collision.

Mr Tompkins noticed, however, that there was a maximum flow of balls in the direction of the original impact.

'Scattering of S-wave,' said a familiar voice behind him, and Mr Tompkins recognized the professor. 'Now,' exclaimed

The white ball went in all directions

Mr Tompkins, 'is there something curved again here? The table seems to me perfectly flat.'

'That is quite correct,' answered the professor; 'space here is quite flat and what you observe is actually a quantum-mechanical phenomenon.'

'Oh, the matrix!' ventured Mr Tompkins sarcastically.

'Or, rather, the uncertainty of motion,' said the professor.

'The owner of the billiard room has collected here several objects which suffer, if I may so express myself, from "quantum-elephantism". Actually all bodies in nature are subject to quantum laws, but the so-called quantum constant which governs these phenomena is very, very small; in fact, its numerical value has twenty-seven zeros after the decimal point. For these balls here, however, this constant is much larger—about unity—and you may easily see with your own eyes phenomena which science succeeded in discovering only by using very sensitive and sophisticated methods of observation.' Here the professor became thoughtful for a moment.

'I do not mean to criticize,' he continued, 'but I *would* like to know where the man got these balls from. Strictly speaking, they could not exist in our world, as, for all bodies in our world, the quantum constant has the same small value.'

'Maybe he imported them from some other world,' proposed Mr Tompkins; but the professor was not satisfied and remained suspicious. 'You have noticed,' he continued, 'that the balls "spread out". This means that their position on the table is not quite definite. You cannot actually indicate the position of a ball exactly; the best you can say is that the ball is "mostly here" and "partially somewhere else".'

'This is very unusual,' murmured Mr Tompkins.

'On the contrary,' insisted the professor, 'it is absolutely usual, in the sense that it is always happening to any material body. Only, owing to the small value of the quantum constant and to the roughness of the ordinary methods of observation, people do not notice this indeterminacy. They arrive at the erroneous conclusion that position or velocity are always definite quantities. Actually both are always indefinite to some extent, and the better one is defined the more the other is spread out. The quantum constant just governs the relation between these two uncertainties. Look here, I am going to put definite limits on the position of this ball by putting it inside a wooden triangle.'

As soon as the ball was placed in the enclosure the whole inside of the triangle became filled up with the glittering of ivory.

'You see!' said the professor, 'I defined the position of the ball to the extent of the dimensions of the triangle, i.e. several inches. This results in considerable uncertainty in the velocity, and the ball is moving rapidly inside the boundary.'

'Can't you stop it?' asked Mr Tompkins.

'No—it is physically impossible. Any body in an enclosed space possesses a certain motion—we physicists call it zero-point motion. Such as, for example, the motion of electrons in any atom.'

While Mr Tompkins was watching the ball dashing to and fro in its enclosure like a tiger in a cage, something very unusual happened. The ball just 'leaked out' through the wall of the triangle and next moment was rolling towards a distant corner of the table. The strange thing was that it really did not jump over the wooden wall, but just passed through it, not rising from the table.

'Well, there you are,' said Mr Tompkins, 'your "zero-motion" has run away. Is that according to the rules?'

'Of course it is,' said the professor, 'in fact this is one of the most interesting consequences of quantum theory. It is impossible to hold anything inside an enclosure provided there is enough energy for running away after crossing the wall. Sooner or later the object will just "leak through" and get away.' 'Then I will never go to the Zoo again,' said Mr Tompkins decisively, and his vivid imagination immediately drew a frightful picture of lions and tigers 'leaking through' the walls of their cages. Then his thoughts took a somewhat different direction: he thought about a car locked safely in a garage leaking out, just like a good old ghost of the middle ages, through the wall of the garage.

'How long have I to wait,' he asked the professor, 'until a car, not made from this kind of stuff here, but just made of ordinary steel, will "leak out" through the wall of, let us say, a brick garage? I would very much like to see that!'

Just like a good old ghost of the middle ages

After making some rapid calculations in his head, the professor was ready with the answer: 'It will take about 1,000,000,000... 000,000 years.'

Even though he was accustomed to large numbers in the bank accounts, Mr Tompkins lost the number of noughts mentioned by the professor—it was, however, long enough for him not to worry about his car running away.

'Suppose I believe all you say. I cannot see, however, how such things could be observed—provided we do not have these balls here.'

'A reasonable objection,' said the professor. 'Of course I do not mean that the quantum phenomena could be observed with such big bodies as those with which you are usually dealing. But the point is that the effects of the quantum laws become much more noticeable in their application to very small masses such as atoms or electrons. For these particles, the quantum effects are so large that ordinary mechanics become quite inapplicable. The collision between two atoms looks exactly like the collision between two balls which you have just observed, and the motion of electrons within an atom resembles very closely the "zero-point motion" of the billiard ball I put inside the wooden triangle.'

'And do the atoms run out of the garage very often?' asked Mr Tompkins.

'Oh yes, they do. You have heard, of course, about radioactive bodies, the atoms of which spontaneously disintegrate, emitting very fast particles. Such an atom, or rather its central part called the atomic nucleus, is quite analogous to a garage in which the cars, i.e. the other particles, are stored. And they do escape by leaking through the walls of this nucleus—sometimes they will not stay inside for a second. In these nuclei, the quantum phenomena become quite usual!'

Mr Tompkins felt very tired after this long conversation and was looking round distractedly. His attention was drawn to a large grandfather clock standing in the corner of the room. The long old-fashioned pendulum was slowly swinging to and fro.

'I see you are interested in this clock,' said the professor. 'This is also a mechanism which is not quite usual—but at present it is out of date. The clock just represents the way people used first to think about quantum phenomena. Its pendulum is arranged in such a way that its amplitude can increase only by finite steps. Now, however, all clockmakers prefer to use the patent spreading-out-pendulums.'

'Oh, I wish I could understand all these complicated things!' exclaimed Mr Tompkins.

'Very well,' retorted the professor, 'I dropped into this pub on the way to my lecture about the quantum theory because I saw you through the window. Now is just the time for me to go, in order not to be late for my lecture. Do you care to come along?'

'Oh yes, I do!' said Mr Tompkins.

As usual the large auditorium was packed with students, and Mr Tompkins was happy even to get a seat on the steps.

Ladies and Gentlemen—began the professor—

In my two previous lectures I tried to show you how the discovery of the upper limit for all physical velocities and the analysis of the notion of a straight line brought us to a complete reconstruction of the classical ideas about space and time.

This development of the critical analysis of the foundations of physics did not, however, stop at this stage, and still more striking discoveries and conclusions have been in store. I am referring to the branch of physics known as quantum theory which is not so much concerned with the properties of space and time themselves as with the mutual interactions and motions of material objects in space and time. In classical physics it was always accepted as self-evident that the interaction between any two physical bodies could be made as small as is required by the conditions of the experiment, and practically reduced to zero whenever necessary. For example, if in investigating the heat developed in certain processes one was afraid that the introduction of a thermometer would take away a certain amount of heat and thus introduce a disturbance in the normal course of the process observed, the experimenter was always certain that by using a smaller thermometer, or a very tiny thermocouple, this disturbance could be reduced to a point below the limits of needed accuracy.

The conviction that any physical process can, in principle, be observed with any required degree of accuracy, without disturbing it by the observation, was so strong that nobody troubled to formulate such a proposition explicitly, and all problems of this kind have always been treated as purely technical difficulties. However, new empirical facts accumulated since the beginning of the present century were steadily bringing physicists to the conclusion that the situation is really much more complicated and that *there exists in nature a certain lower limit of interaction which can never be surpassed.* This natural limit of accuracy is negligibly small for all kinds of processes with which we are familiar in ordinary life, but it becomes quite important when we are handling the interactions taking place in such tiny mechanical systems as atoms and molecules.

In the year 1900 the German physicist MAX PLANCK, while investigating theoretically the conditions of equilibrium between matter and radiation, came to the surprising conclusion that no such equilibrium is possible unless we suppose that *the interaction between the matter and radiation takes place not continuously, as we always supposed, but in a sequence of separate 'shocks'*, a definite amount of energy being transferred from matter to radiation or vice versa in each of these elementary acts of interaction. In order to get the desired equilibrium, and to achieve agreement with the experimental facts, it was necessary to introduce a simple mathematical relation of proportionality between the amount of energy transferred in each shock and the frequency (inverse period) of the process leading to the transfer of energy.

Thus, denoting the coefficient of proportionality by a symbol 'h' Planck was forced to accept that the minimal portion, or quantum, of energy transferred must be given by the expression

$$E = h\nu, \tag{1}$$

where ν stands for frequency. The constant h has the numerical value $6{\cdot}547 \times 10^{-27}$ ergs $\times$ second and is usually called Planck's constant or the quantum constant. Its small numerical value is responsible for the fact that quantum phenomena are usually not observed in our everyday life.

The further development of Planck's ideas is due to Einstein who, a few years later, came to the conclusion that *not only is the radiation emitted in definite discrete portions, but that it always exists in this way, consisting of a number of discrete 'packages of energy' which he called light quanta.*

In so far as light quanta are moving they should possess, apart from their energy $h\nu$, a certain mechanical momentum also, which, according to relativistic mechanics, should be equal to their energy divided by the velocity of light c. Remembering that the frequency of light is related to its wave length λ by the relation $\nu = c/\lambda$, we can write for the mechanical momentum of a light quantum:

$$p = \frac{h\nu}{c} = \frac{h}{\lambda}. \tag{2}$$

Since the mechanical action produced by the impact of a moving object is given by its momentum we must conclude that the action of light quanta increases with their decreasing wave length.

One of the best experimental proofs of the correctness of the idea of light quanta, and the energy and momentum ascribed to them, was given by the investigation of the American physicist ARTHUR COMPTON who, studying the collisions between light quanta and electrons, arrived at the result that electrons set into motion by the action of a ray of light behaved exactly as if they had been struck by a particle with the energy and momentum given by the previously given formulae. The light quanta themselves, after the collision with electrons, were also shown to suffer certain changes (in their frequency), in excellent agreement with the prediction of the theory.

We can say at present that, as far as the interaction with matter is concerned, the quantum property of radiation is a well established experimental fact.

The further development of the quantum ideas is due to the

famous Danish physicist NIELS BOHR who, in 1913, was first to express the idea that *the internal motion of any mechanical system may possess only a discrete set of possible energy values and the motion can change its state only by finite steps*, a definite amount of energy being radiated in each of such transitions. The mathematical rules defining the possible states of mechanical systems are more complicated than in the case of radiation and we will not enter here into their formulation. We shall only indicate that, just as, in the case of light quanta, the momentum is defined through the wave length of light, so in the mechanical system the momentum of any moving particle is connected with the geometrical dimensions of the region of space in which it is moving, its order of magnitude being given by the expression

$$p_{\text{particle}} \cong \frac{h}{l}, \tag{3}$$

l being here linear dimensions of the region of motion. Due to the extremely small value of the quantum constant quantum phenomena could be of importance only for motions taking place in such small regions as the inside of atoms and molecules, and they play a very important part in our knowledge of the internal structure of matter.

One of the most direct proofs of the existence of the sequence of discrete states of these tiny mechanical systems was given by the experiments of JAMES FRANCK and GUSTAV HERTZ who, bombarding atoms by electrons of varying energy, noticed that definite changes in the state of the atom took place only when the energy of the bombarding electrons reached certain discrete values. If the energy of electrons was brought below a certain limit no effect whatsoever was observed in the atoms because the amount of energy carried by each electron was not enough to raise the atom from the first quantum state into the second.

Thus at the end of this first preliminary stage of the development of quantum theory the situation could be described, not as the modification of fundamental notions and principles of classical physics, but as its more or less artificial restriction by rather mysterious quantum conditions picking out from the continuous variety of classical possible motions only a discrete set of 'permitted' ones. If, however, we look deeper into the connexion between the laws of classical mechanics and these quantum conditions required by our extended experience, we shall discover that the system obtained by their unification suffers from logical inconsistency, and that the empirical quantum restrictions make senseless the fundamental notions on which classical mechanics is based. In fact, the fundamental concept concerning motion in classical theory is that any moving particle occupies at any given moment a certain position in space and possesses a definite velocity characterizing the time changes of its position on the trajectory.

These fundamental notions of position, velocity, and trajectory, on which are based all the elaborated building of classical mechanics, are formed (as are all our other notions) on observation of the phenomena around us, and, like the classical notions of space and time, might be subject to far reaching modifications as soon as our experience extends into new, previously unexplored, regions.

If I ask somebody why he believes that any moving particle occupies at any given moment a certain position describing in the course of time a definite line called the trajectory, he will most probably answer: 'Because I see it this way, when I observe the motion.' Let us analyse this method of forming the classical notion of the trajectory and see if it really will lead to a definite result. For this purpose we imagine a physicist supplied with any kind of the most sensitive apparatus, trying to pursue the motion of a little material body thrown from the wall of his laboratory. He decides to make his observation by 'seeing' how the body moves and for this purpose he uses a small but very precise theodolite. Of course to see the moving body he must illuminate it and, knowing that light in general produces a pressure on the body and might disturb its motion, he decides to use short flash illumination only at the moments when he makes the observation. For his first trial he wants to observe only ten points on the trajectory and thus he chooses his flashlight source so weak that

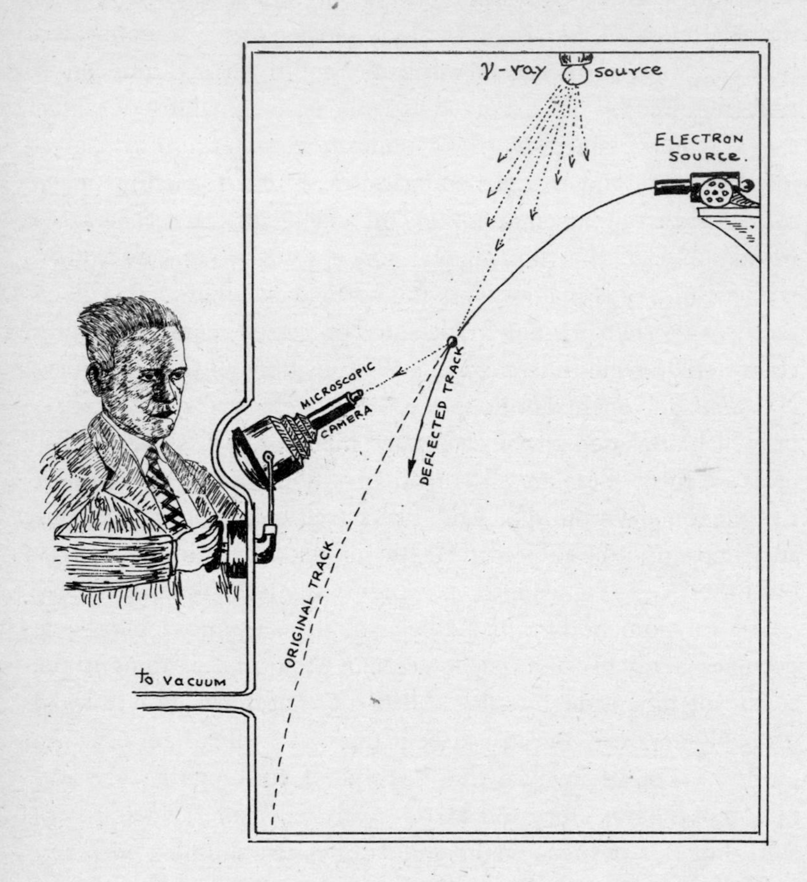

Heisenberg's γ-ray microscope

the integral effect of light pressure during ten successive illuminations should be within the accuracy he needs. Thus, flashing his light ten times during the fall of the body, he obtains, with the desired accuracy, ten points on the trajectory.

Now he wants to repeat the experiment and to get one hundred points. He knows that a hundred successive illuminations will disturb the motion too much and therefore, preparing for the second set of observations, chooses his flashlight ten times less intense. For the third set of observations, desiring to have one thousand points, he makes the flashlight a hundred times fainter than originally.

Proceeding in this way, and constantly decreasing the intensity of his illumination, he can obtain as many points on the trajectory as he wants to, without increasing the possible error above the limit he had chosen at the beginning. This highly idealized, but in principle quite possible, procedure represents the strictly logical way to construct the motion of a trajectory by 'looking at the moving body' and you see that, in the frame of classical physics, it is quite possible.

But now let us see what happens if we introduce the quantum limitations and take into account the fact that the action of any radiation can be transferred only in the form of light quanta. We have seen that our observer was constantly reducing the amount of light illuminating the moving body and we must now expect that he will find it impossible to continue to do so as soon as he comes down to one quantum. Either all or none of the total light quantum will be reflected from the moving body, and in the latter case the observation cannot be made. Of course we have seen that the effect of collision with a light quantum decreases with increasing wave length, and our observer, knowing it too, will certainly try to use for his observations light of increasing wave length to compensate for the number of observations. But here he will meet with another difficulty.

It is well known that when using light of certain wave lengths one cannot see details smaller than the wave length used; in fact one cannot paint a Persian miniature using a house-painter's brush! Thus, by using longer and longer waves, he will spoil the estimate of each single point and soon will come to the stage where each estimate will be uncertain by an amount comparable to the size of all his laboratory and more. Thus he will be forced finally to a compromise between the large number of observed points and uncertainty of each estimate and will never be able to arrive at an

Little bells on springs

exact trajectory as a mathematical line such as that obtained by his classical colleagues. His best result will be a rather broad washed-out band, and if he bases his notion of the trajectory on the result of his experience, it will be rather different from a classical one.

The method discussed here is an optical method, and we can now try another possibility, using a mechanical method. For this purpose our experimenter can devise some tiny mechanical apparatus, say little bells on springs, which would register the passage of material bodies if such a body passes close to them. He can spread a large number of such 'bells' through the space through which the moving body is expected to pass and after the passage the 'ringing of bells' will indicate its track. In classical physics one can make the 'bells' as small and sensitive as one likes and, in the limiting case of an infinite number of infinitely small bells, the notion of a trajectory can be again formed with any desired accuracy. However, the quantum limitations for mechanical systems will spoil the situation again. If the 'bells' are too small, the amount of momentum which they will take from the moving body will be, according to formula (3), too large and the motion will be largely disturbed even after only one bell has been hit. If the bells are large the uncertainty of each position will be very large. The final trajectory deduced will again be a spread-out band!

I am afraid that all these considerations about an experimenter trying to observe the trajectory may make a somewhat too technical impression, and you will be inclined to think that, even if our observer cannot estimate the trajectory by the means he is using, some other more complicated device will give the desired result. I must remind you, however, that we have here been discussing not any particular experiment done in some physical laboratory, but an idealization of the most general question of physical measurement. As far as any actions existing in our world can be classified either as due to radiative field or as purely mechanical, any elaborated scheme of measurement will be necessarily reduced to the elements described in these two methods and will finally lead to the same result. As far as our ideal 'measuring apparatus' can involve all the physical world we should come ultimately to the conclusion that such things as exact position and a trajectory of precise shape have no place in a world subject to quantum laws.

Let us now return to our experimenter and try to get the mathematical form for the limitations imposed by quantum conditions. We have already seen that in both methods used there is always a conflict between the estimate of position and the disturbance of the velocity of the moving object. In the optical method, the collision with a light quantum will, because of the mechanical law of conservation of momentum, introduce an uncertainty in the momentum of the particle comparable with the momentum of the light quantum used. Thus, using formula (2), we can write for the uncertainty of momentum of the particle

$$\Delta p_{\text{particle}} \cong \frac{h}{\lambda}, \tag{4}$$

and remembering that the uncertainty of position of the particle is given by the wave length ($\Delta q \cong \lambda$) we deduce:

$$\Delta p_{\text{particle}} \times \Delta q_{\text{particle}} \cong h. \tag{5}$$

In the mechanical method the momentum of the moving particle will be made uncertain by the amount taken by the 'bells'. Using our formula (3) and remembering that in this case the uncertainty of position is given by the size of the bell ($\Delta q \cong l$), we come again to the same finite formula as in the previous case. Thus the relation (5), first formulated by the German physicist WERNER HEISENBERG, represents the fundamental uncertainty—relation of quantum theory—*the better one defines the position, the more indefinite the momentum becomes, and vice versa.*

Remembering that momentum is the product of the mass of the moving particle and its velocity, we can write

$$\Delta v_{\text{particle}} \times \Delta q_{\text{particle}} \cong \frac{h}{m_{\text{particle}}}. \tag{6}$$

For bodies which we usually handle this is ridiculously small. For a lighter particle of dust with the mass 0·000,000,1 gm both position and velocity can be measured with an accuracy of 0·000,000,01 %! However, for an electron (with the mass 10^{-29} gm) the product $\Delta v \Delta q$ should be of the order 100. Inside an atom the velocity of an electron should be defined at least within $\pm 10^{10}$ cm/sec otherwise it will escape from the atom. This gives for the uncertainty of position 10^{-8} cm, i.e. the total dimensions of an atom. Thus 'the orbit' of an electron in an atom is spread out by such extent that 'the thickness' of the trajectory becomes equal to its 'radius'. *Thus the electron appears simultaneously all around the nucleus.*

During the last twenty minutes I have tried to show you a picture of the disastrous results of our criticism of classical ideas of motion. The elegant and sharply defined classical notions are broken to pieces and give place to what I would call a shapeless gruel. You may naturally ask me how on earth the physicists are going to describe any phenomena in view of this ocean of uncertainty. The answer is that we have so far destroyed classical notions, but we have not yet arrived at an exact formulation of new ones.

We shall proceed with it now. It is clear that, if we cannot in general define the position of a material particle by a mathematical point and the trajectory of its motion by a mathematical line because the things had spread out, we should use other methods of description giving, so to speak, 'the density of the gruel' at different points of space. Mathematically it means the use of continuous functions (such as are used in hydrodynamics) and physically this requires us to be used to the expressions like 'this object is mostly here, but partially there and even yonder', or 'this coin is 75 % in my pocket and 25 % in yours'. I know that such sentences will terrify you, but, due to the small value of the quantum constant, you will never need them in everyday life. However, if you are going to study atomic physics, I would strongly advise you to get accustomed to such expressions first.

I must warn you here against the erroneous idea that the function describing the 'density of presence' has a physical reality in our ordinary three-dimensional space. In fact, if we describe the behaviour of, say, two particles, we must answer the question concerning the presence of our first particle in one place and the simultaneous presence of our second particle in some other place; to do this we have to use a function of six variables (coordinates of two particles) which cannot be 'localized' in three-dimensional space. For more complex systems functions of still larger numbers of variables must be used. In this sense, the 'quantum mechanical function' is analogous to the 'potential function' of a system of particles in classical mechanics or to the 'entropy' of a system in statistical mechanics. It only *describes* the motion and helps us to predict the result of any particular motion under given conditions. The physical reality stays with the particles the motion of which we are describing.

The function which describes to what extent the particle or system of particles is present in different places requires some mathematical notation and, according to the Austrian physicist ERWIN SCHRÖDINGER, who first wrote the equation defining the behaviour of this function, it is denoted by the symbol '$\psi\bar{\psi}$'.

I am not going to enter here into the mathematical proof of his fundamental equation, but I will draw your attention to the requirements which lead to its derivation. The most important of these requirements is a very unusual one: *The equation must be written in such a way that the function which describes the motion of material particles should show all the characteristics of a wave.*

The necessity of ascribing wave properties to the motion of material particles was first indicated by the French physicist LOUIS DE BROGLIE, on the basis of his theoretical studies of the structure of an atom. In the following years the wave properties of the motion of material particles were firmly established by numerous experiments, showing such phenomena as the *diffraction* of a beam of electrons passing through a small opening and *interference phenomena* taking place even for such comparatively large and complex particles as molecules.

The observed wave properties of material particles were absolutely incomprehensible from the point of view of classical conceptions of motion, and de Broglie himself was forced to a rather unnatural point of view: that the particles are 'accompanied' by certain waves which, so to speak, 'direct' their motion.

However, as soon as the classical notions are destroyed and we come to the description of motion by continuous functions, the requirement of wave character becomes much more understandable. It just says that the propagation of our '$\psi\bar{\psi}$' function is not analogous to (let us say) propagation of heat through a wall heated on one side but rather to the propagation of mechanical deformation (sound) through the same wall. Mathematically it requires a definite rather restricted form of the equation we are looking for. This fundamental condition, together with the additional requirement that our equations should go over into the equations of classical mechanics when applied to particles of large mass for which quantum effect should become negligible, practically reduces the problem of finding the equation to a purely mathematical exercise.

If you are interested in how the equation looks in its final form, I can write it here for you. Here it is:

$$\nabla^2\psi + \frac{4\pi mi}{h}\dot{\psi} - \frac{8\pi^2 m}{h}U\psi = 0. \qquad (7)$$

In this equation the function U represents the potential of forces acting on our particles (with the mass m), and it gives a definite solution of the problem of motion for any given distribution of force. The application of this 'Schrödinger's wave equation' has allowed physicists, during the forty years of its existence, to develop the most complete and logically consistent picture of all phenomena taking place in the world of atoms.

Some of you may have been wondering that until now I have not used the word 'matrix', often heard in connexion with the quantum theory. I must confess that personally I rather dislike these matrices and prefer to do without them. But, in order not to leave you absolutely ignorant about this mathematical implement of the quantum theory, I shall say a word or two about it. The motion of a particle or of a complex mechanical system is always described, as you have seen, by certain continuous wave functions. These functions are often rather complicated and can be represented as being composed of a number of simpler oscillations, the so-called 'proper functions', much in the way that a complicated sound can be made up from a number of simple harmonic notes.

One can describe the whole complex motion by giving the amplitudes of its different components. Since the number of components (overtones) is infinite we must write infinite tables of amplitudes in a form:

$$\begin{matrix} q_{11} & q_{12} & q_{13} & \cdots \\ q_{21} & q_{22} & q_{23} & \cdots \\ q_{31} & q_{32} & q_{33} & \cdots \\ \cdots & \cdots & \cdots & \cdots \end{matrix} \qquad (8)$$

Such a table, which is subject to comparatively simple rules of mathematical operations, is called a 'matrix' corresponding to a given motion, and some theoretical physicists prefer to operate with matrices instead of dealing with the wave functions themselves. Thus the 'matrix mechanics' as they sometimes call it is just a mathematical modification of the ordinary 'wave mechanics'; and in these lectures, devoted mainly to the principal questions, we do not need to enter more deeply into these problems.

I am very sorry that time does not permit me to describe to you the further progress of quantum theory in its relation to the theory of relativity. This development, due mainly to the work of the British physicist PAUL ADRIEN MAURICE DIRAC, brings in a number of very interesting points and has also led to some extremely important experimental discoveries. I may be able to return at some other time to these problems, but here at present I must stop, and express the hope that this series of lectures has helped you to get a clearer picture of the present conception of the physical world and has excited in you an interest for further studies.

8

Quantum Jungles

Next morning Mr Tompkins was dozing in bed, when he became aware of somebody's presence in the room. Looking round, he discovered that his old friend the professor was sitting in the armchair, absorbed in the study of a map spread on his knee.

'Are you coming along?' asked the professor, lifting his head.

'Coming where?' said Mr Tompkins, still wondering how the professor had got into his room.

'To see the elephants, of course, and the rest of the animals of the quantum jungle. The owner of the billiard room we visited recently told me his secret about the place where the ivory for his billiard-balls came from. You see this region which I've marked with red pencil on the map? It seems that everything within it is subject to quantum laws with a very large quantum constant. The natives think that all this part of the country is populated by devils, and I am afraid it will hardly be possible for us to find a guide. But if you want to come along, you had better hurry up. The boat is sailing in an hour's time and we still have to pick up Sir Richard on our way.'

'Who is Sir Richard?' asked Mr Tompkins.

'Haven't you ever heard about him?' The professor was evidently surprised. 'He is a famous tiger-hunter, and decided to go with us, when I promised him some interesting shooting.'

They came to the docks just in time to see the loading of a number of long boxes containing Sir Richard's rifles and the special bullets made from lead which the professor had obtained from the lead mines near the quantum jungle. While Mr Tompkins was arranging his baggage in the cabin, the steady vibrations of the boat told him that they were off. The sea journey was nothing remarkable, and Mr Tompkins scarcely noticed the time until they came ashore in a fascinating oriental city, the nearest populated place to the mysterious quantum regions.

'Now,' said the professor, 'we have to buy an elephant for our journey inland. As I do not think any of the natives will agree to go with us, we shall have to drive the elephant ourselves, and you, my dear Tompkins, will have to learn the job. I shall be too busy with my scientific observations and Sir Richard will have to handle the firearms.'

Mr Tompkins was rather unhappy when, coming to the elephant market on the outskirts of the city, he saw the huge animals, one of which he would have to handle. Sir Richard, who knew a lot about elephants, picked out a nice big animal and asked the owner what price it was.

'Hrup hanweck 'o hobot hum. Hagori ho, haraham oh Hohohohi,' said the native, showing his shining teeth.

'He wants quite a lot of money for it,' translated Sir Richard, 'but says that this is an elephant from the quantum jungle and it is therefore more expensive. Shall we take it?'

'By all means,' explained the professor. 'I heard on the boat that sometimes elephants come from the quantum lands and are caught by the natives. They are much better than the elephants from other regions, and in our case we shall have quite an advantage because this animal will feel at home in the jungle.'

Mr Tompkins inspected the elephant from all sides; it was a very beautiful, large animal, but there was no marked difference in its behaviour from the elephants he had seen in the Zoo. He turned to the professor—'You said that this was a quantum elephant, but it looks just like an ordinary elephant to me, and does not behave in a funny way, like the billiard-balls made from the tusks of some of its relatives. Why doesn't it spread out in all directions?'

'You show a peculiar slowness of comprehension,' said the professor. 'It is because of its very large mass. I told you some time ago that all the uncertainty in position and velocity depends on the mass; the larger the mass, the smaller the uncertainty. That is why the quantum laws have not been observed in the ordinary world even for such light bodies as particles of dust, but become quite important for electrons, which are billions of billions of times lighter. Now, in the quantum jungle, the quantum constant is rather large, but still not large enough to produce striking effects in the behaviour of such a heavy animal as an elephant. The uncertainty of the position of a quantum elephant can be noticed only by close inspection of its contours. You may have noticed that the surface of its skin is not quite definite and seems to be slightly fuzzy. In course of time this uncertainty increases very slowly, and I think this is the origin of the native legend that very old elephants from the quantum jungle possess long fur. But I expect that all smaller animals will show very remarkable quantum effects.'

'Isn't it nice,' thought Mr Tompkins, 'that we are not doing this expedition on horseback? If that were the case, I should probably never know whether my horse was between my knees or in the next valley.'

After the professor and Sir Richard with his rifles had climbed into the basket fastened on to the elephant's back, and Mr Tompkins, in his new capacity of mahout, had taken his position on the elephant's neck, clutching the goad in one hand, they started towards the mysterious jungle.

The people in the city told them that it would take about an hour to get there, and Mr Tompkins, trying to keep his balance between the elephant's ears, decided to make use of the time by learning more about quantum phenomena from the professor.

'Can you tell me, please,' he asked, turning to the professor, '*why* do bodies with small mass behave so peculiarly, and what is the commonsense meaning of this quantum constant that you are always talking about?'

'Oh, it is not so difficult to understand,' said the professor. 'The funny behaviour of all objects you observe in the quantum world is just due to the fact that you are looking at them.'

'Are they so shy?' smiled Mr Tompkins.

'"Shy" is an unsuitable word,' said the professor bleakly. 'The point is, however, that in making any observation of the motion you will necessarily disturb this motion. In fact, if you learn something about the motion of a body, this means that the moving body delivered some action on your senses or the apparatus you are using. Owing to the equality of action and reaction we must conclude that your measuring apparatus also acted on the body and, so to speak, "spoiled" its motion, introducing an uncertainty in its position and velocity.'

'Well,' said Mr Tompkins, 'if I had touched that ball in the billiard room with my finger I should certainly have disturbed its motion. But I was just looking at it; does that disturb it?'

'Of course it does. You cannot see the ball in darkness, but if you put on the light, the light-rays reflected from the ball and making it visible will act on the ball—light-pressure we call it—and "spoil" its motion.'

'But suppose I used very fine and sensitive instruments, can't I make the action of my instruments on the moving body so small as to be negligible?'

'That is just what we thought in classical physics, before the *quantum of action* was discovered. At the beginning of this century it became clear that the *action* on any object cannot be brought below a certain limit which is called the quantum constant and usually denoted by the symbol "*h*". In the ordinary world the quantum of action is very small; in customary units it is expressed by a number with twenty-seven zeros after the decimal point, and is of importance only for such light particles as electrons which, owing to their very small mass, will be influenced by very small actions. In the quantum jungle we are now approaching, the quantum of action is very big. This is a rough world where no gentle action is possible. If a person in such a world tried to pet a kitten, it would either not feel anything at all, or its neck would be broken by the first quantum of caress.'

'This is all very well,' said Mr Tompkins thoughtfully, 'but when nobody is looking, do the bodies behave properly, I mean, in the way we are accustomed to think?'

'When nobody is looking,' said the professor, 'nobody can know how they do behave, and thus your question has no physical sense.'

'Well, well,' exclaimed Mr Tompkins, 'it certainly looks like philosophy to me!'

'You can call it philosophy if you like'—the professor was evidently offended—'but as a matter of fact, this is the fundamental principle of modern physics—*never to speak about the things you cannot know*. All modern physical theory is based on this principle, whereas the philosophers usually overlook it. For example, the famous German philosopher KANT spent quite a lot of time reflecting about the properties of bodies not as they "appear to us", but as they "are in themselves". For the modern physicist only the so-called "observables" (i.e. principally, observable properties) have any significance, and all modern physics is based on their mutual relation. The things which cannot be observed are good only for idle thinking—you have no restrictions in inventing them, and no possibility of checking their existence, or of making any use of them. I should say....'

At this moment a terrible roar filled the air and their elephant jerked so violently that Mr Tompkins almost fell off. A large pack of tigers was attacking their elephant, jumping simultaneously from all sides. Sir Richard grabbed his rifle and pulled the trigger, aiming right between the eyes of the tiger nearest to him. The next moment Mr Tompkins heard him mutter a strong expression common among hunters; he shot right through the tiger's head without causing any damage to the animal.

'Shoot more!' shouted the professor. 'Scatter your fire all round and don't mind about precise aiming! There is only one tiger, but it is spread around our elephant and our only hope is to raise the Hamiltonian.'

The professor grabbed another rifle and the cannonade of shooting became mixed up with the roar of the quantum tiger. An eternity passed, so it seemed to Mr Tompkins, before all was over. One of the bullets 'hit the spot' and, to his great surprise, the tiger, which became suddenly one, was vigorously hurled away, its dead body describing an arc in the air, and landing somewhere behind the distant palm grove.

'Who is this Hamiltonian?' asked Mr Tompkins after things had quietened down. 'Is he some famous hunter you wanted to raise from the grave to help us?'

'Oh!' said the professor, 'I am so sorry. In the excitement of battle I started to use scientific language—which you cannot understand! Hamiltonian is a mathematical expression describing the quantum interaction between two bodies. It is named after an Irish mathematician, HAMILTON, who first used this mathematical form. I just wanted to say that by shooting more quantum bullets we increase the probability of the interaction between the bullet and the body of the tiger. In the quantum world, you see, one cannot aim precisely and be sure of a hit. Owing to the spreading out of the bullet, and of the aim itself, there is always only a finite chance of hitting, never a certainty. In our case we fired at least thirty bullets before we actually hit the tiger; and then the action of the bullet on the tiger was so violent that it hurled its body far away. The same things are happening in our world at home but on a much smaller scale. As I have already mentioned, in the ordinary world one has to investigate the behaviour of such small particles as electrons in order to notice anything. You may have heard that each atom consists of a comparatively heavy nucleus and a number of electrons rotating round it. One used to think, at first, that the motion of electrons round the nucleus is quite analogous to the motion of planets round the sun, but deeper analysis has shown that ordinary notions concerning the motion are too rough for such a miniature system as that of the atom. The actions which play an important role inside an atom are of the same order of

A large pack of fuzzy-looking tigers was attacking their elephant magnitude as the elementary quantum of action and thus the whole picture is largely spread out. The motion of the electron round the atomic nucleus is in many respects analogous to the motion of our tiger, which seemed to be all round the elephant.'

'And does somebody shoot at the electron as we did the tiger?' asked Mr Tompkins.

'Oh yes, of course, the nucleus itself sometimes emits very energetic light quanta or elementary action-units of light. You can also shoot at the electron from outside the atom, by illuminating it with a beam of light. And it all happens there just as with our tiger here: many light quanta pass through the location of the electron without affecting it, until presently one of them acts on the electron and throws it out of the atom. The quantum system cannot be affected slightly; it is either not affected at all, or else changed a lot.'

'Just as with the poor kitten which cannot be petted in the quantum world without being killed,' concluded Mr Tompkins.

'Look! gazelles, and lots of them!' exclaimed Sir Richard, raising his rifle. In fact a big herd of gazelles was emerging from the bamboo grove.

'Trained gazelles,' thought Mr Tompkins. 'They run in as regular formation as soldiers on parade. I wonder if this is also some quantum effect.'

The group of gazelles which was approaching their elephant was moving rapidly and Sir Richard was ready to shoot, when the professor stopped him.

'Do not waste your cartridges,' he said, 'there is very little chance of hitting an animal when it is moving in a diffraction pattern.'

'What do you mean by "*an*" animal?' exclaimed Sir Richard. 'There are at least several dozens of them!'

'Oh no! There is only one little gazelle which, because it is scared of something, is running through the bamboo grove. Now, the "spread-out" of all bodies possesses a property analogous to that of ordinary light; and, passing through a regular sequence of

Sir Richard was ready to shoot, when the professor stopped him

openings, for instance between the separate bamboo trunks in the grove, it shows the phenomena of diffraction about which you might have heard at school. We speak therefore about the wave-character of matter.'

But neither Sir Richard nor Mr Tompkins could think at all what this mysterious word 'diffraction' might mean, and the conversation stopped at this point.

Passing farther through the quantum land our travellers met quite a lot of other interesting phenomena, such as quantum mosquitoes, which could scarcely be located at all, owing to their small mass, and some very amusing quantum monkeys. Now they were approaching something which looked very much like a native village.

'I did not know,' said the professor, 'that there was a human population in these regions. Judging by the noise, I suppose they are having some sort of festival. Listen to this incessant noise of bells.'

It was very difficult to distinguish the separate figures of natives who were evidently dancing a wild dance round the big fire. Brown hands with bells of all sizes were constantly rising from among the crowd. As they approached still closer, everything, including the huts and surrounding big trees, began to spread out, and the ringing of the bells became unbearable to Mr Tompkins's ears. He stretched his hand out, grabbed something, and then threw it away. The alarm clock hit the glass of water standing on his night-table and the cold stream of water brought him to his senses. He jumped up, and started to dress rapidly. In half an hour he must be at the bank.

9

Maxwell's Demon

During many months of unusual adventures, in the course of which the professor tried to introduce Mr Tompkins to the secrets of physics, Mr Tompkins became more and more enchanted by Maud and finally, and rather sheepishly, made a proposal of marriage. This was readily accepted, and they became man and wife. In his new role of father-in-law, the professor considered it his duty to enlarge the knowledge of his daughter's husband in the field of physics and of its most recent progress.

One Sunday afternoon Mr and Mrs Tompkins were resting in armchairs in their comfortable flat, she being engulfed in the latest issue of *Vogue*, he reading an article in *Esquire*.*

'Oh,' Mr Tompkins exclaimed suddenly, 'here is a chance game system which really works!'

'Do you really think, Cyril, that it will?' asked Maud, raising her eyes reluctantly from the pages of the fashion magazine. 'Father has always said that there can't be such a thing as a sure-fire gambling system.'

'But look here, Maud,' answered Mr Tompkins, showing her the article he had been studying for the last half hour. 'I don't know about other systems, but this one is based on pure and simple mathematics, and I really don't see how it could possibly go wrong. All you have to do is to write down three figures

1, 2, 3

on a piece of paper, and follow a few simple rules given here.'

'Well, let's try it out,' suggested Maud, beginning to be interested. 'What are the rules?'

'Suppose you follow the example given in the article. That's

* January 1940.

'But you *must* win this time!'

probably the best way to learn them. As illustration, they have used a roulette game in which you place your money on red or black, which is the same as betting heads or tails on the flip of a coin. I write down

1, 2, 3

and the rule is that my bet must always be the sum of the outside figures in the series. So I take one plus three, which is four, chips and put them, let's say, on red. If I win, I cross out the figures 1 and 3 and my next bet will be the remaining figure 2. If I lose, I add the amount lost to the end of the series and apply the same rule to find my next bet. Well, suppose the ball stops on black and the croupier rakes in my four chips. Then my new series will be

1, 2, 3, 4

and my next bet one plus four, which is five. Suppose I lose a second time. The article says I must keep on in the same way, adding the figure 5 at the end of the series and putting six chips on the table.'

'But you *must* win this time!' cried Maud, getting quite excited. 'You can't keep on losing.'

'Not necessarily,' said Mr Tompkins. 'When I was a boy I used to flip pennies with my friends, and believe it or not, I once saw heads come up ten times in a row. But let's suppose, as this article does, that I win this time. Then I collect twelve chips, but I am still out three chips compared with my original stake. Following the rules, I must cross out the figures 1 and 5, and my series now reads

~~1~~, 2, 3, 4, ~~5~~

My next bet must be two plus four, or six chips again.'

'It says here you have lost again,' sighed Maud, reading over her husband's shoulder. 'That means you have to add six to the series and bet eight chips next time. Isn't that so?'

'Yes, that's right, but I lose again. My series is now

~~1~~, 2, 3, 4, ~~5~~, 6, 8

and I have to bet ten this time. It wins. I cross out the figures 2 and 8 and my next bet is three plus six which is nine. But I lose again.'

'It's a bad example,' said Maud, pouting. 'So far, you've lost three times and only won once. It's not fair!'

'Never mind, never mind,' said Mr Tompkins with the quiet confidence of a magician. 'We'll win all right at the end of the cycle. I lost nine chips on the last spin, so I'll add this figure to the series to make it

~~1~~, ~~2~~, 3, 4, ~~5~~, 6, ~~8~~, 9

and bet twelve chips. I win this time, so I cross out the figures 3 and 9 and bet the sum of the remaining two, or ten chips. The second successive win completes the cycle as all the figures are now crossed out. And I am six chips up, in spite of the fact that I won only four times and lost five!'

'Are you sure you are six chips up?' asked Maud doubtfully.

'Quite sure. You see the system is arranged in such a way that, whenever the cycle is complete, you are always six chips up. You can prove it by simple arithmetic, and that's why I say this system is mathematical and can't fail. If you don't believe it, take a piece of paper and check it yourself.'

'All right. I'll take your word for it that that's the way it works out,' said Maud thoughtfully, 'but, of course, six chips aren't very much to win.'

'Yes they are, if you are sure of winning them at the end of each cycle. You can repeat the procedure over and over again, beginning each time with 1, 2, 3, and making as much money as you want. Isn't it *grand*?'

'Wonderful!' exclaimed Maud. 'Then you can drop your work at the bank, we can move into a better house, and I saw a darling mink coat in a shop window today. It cost only....'

'Of course we'll buy it, but first we had better get to Monte Carlo quickly. A lot of other people must have read this article, and it would be too bad to get there only to find some other fellow had beaten us to it and put the Casino into bankruptcy.'

'I'll ring up the air line,' suggested Maud, 'and find out when the next plane leaves.'

'What's all the hurry about?' said a familiar voice in the hall. Maud's father came into the room and looked at the excited pair in surprise.

'We're leaving for Monte Carlo on the first plane and we're

going to come home very rich,' said Mr Tompkins, rising to greet the professor.

'Oh, I see,' smiled the latter, making himself comfortable in an old-fashioned armchair near the fireplace. 'You have a new gambling system?'

'But this time it's a real one, Father!' protested Maud, her hand still on the phone.

'Yes,' added Mr Tompkins, handing the professor the magazine. 'This one can't miss.'

'Can't it?' said the professor with a smile. 'Well, let's see.' After a short inspection of the article, he went on, 'The distinguishing feature of this system is that the rule governing the amount of your bets calls for you to raise your bet after each loss and, on the other hand, to lower your bet after each win. So, if you should win and lose alternately and with complete regularity, your capital would oscillate up and down, each increase being, however, slightly larger than the previous decrease. In such a case you would, of course, become a millionaire in no time. But as you no doubt understand, such regularity usually does not occur. As a matter of fact, the probability of such a regularly alternating series is just as small as the probability of an equal number of straight wins. So we must see what happens if you have a sequence of several successive wins or losses. If you get what gamblers call a streak of luck, the rule forces you to lower, or at least not to raise, your bet after each win, so your total winnings will not be very high. On the other hand, as you must raise your bet after each loss, a streak of bad luck will be more catastrophic and may throw you out of the game. You can now see that the curve representing the variations in your capital will consist of several slowly rising portions interrupted by very sharp drops. At the beginning of the game, it is likely that you will get on to the long, slowly rising part of the curve and will enjoy for a while the pleasant feeling of watching your money slowly but surely increasing. However, if you go on long enough, in the hope of larger and larger profit, you will come unexpectedly to the sharp drop which might be deep enough to make you bet and lose your last penny. One can show, in a quite general way, that with this or any other system the probability that the curve will reach the double mark is equal to that of reaching zero. In other words, the chances of finally winning are exactly the same as if you put all your money on red or black and double your capital or lose everything on just one spin of the wheel. All that such a system can do is to prolong the game and give you more fun for the money. But if that is all you want to do, you don't have to make it so complicated. There are thirty-six numbers on a roulette wheel, you know, and there is nothing to keep you from covering every number but one. Then the chances are thirty-five out of thirty-six that you will win and that the bank will pay you one chip more than the thirty-five you bet. However, about once in thirty-six spins the ball will stop on the particular number you chose not to cover with a chip, and you will lose all thirty-five. Play this way long enough and the curve of your fluctuating capital will look exactly like the curve you will get by following this magazine's system.

'Of course I have been assuming right along that the bank is taking no cut. As a matter of fact, every roulette wheel I have seen has a zero, and often a double zero as well, which raises the odds against the player. Regardless of the system he uses, therefore, the gambler's money gradually leaks from his pocket to the proprietor's.'

'You mean to say,' said Mr Tompkins dejectedly, 'that there is no such thing as a good gambling system, and that there is no possible way of winning money without risking the slightly higher probability of losing it?'

'That is precisely what I mean,' said the professor. 'What is more, what I have said applies not only to such comparatively unimportant problems as games of chance, but to have great variety of physical phenomena which, at first sight, seem to have nothing to do with the laws of probability. For that matter, if you could devise a system for beating the laws of chance, there are much more exciting things than winning money one could do with it. One could build cars that ran without gasoline, factories that could be operated without coal and plenty of other fantastic things.'

'I've read something somewhere about such hypothetical machines—perpetual motion machines, I believe they are called,' said Mr Tompkins. 'If I remember correctly, machines planned to run without fuel are considered impossible because one cannot manufacture energy out of nothing. Anyway, such machines have no connection with gambling.'

'You are quite right, my boy,' agreed the professor, pleased that his son-in-law knew something at least about physics. 'This kind of perpetual motion, "perpetual motion machines of the first type" as they are called, cannot exist because they would be contrary to the law of the Conservation of Energy. However the fuel-less machines I have in mind are of a rather different type and are usually known as "perpetual motion machines of the second type". They are not designed to create energy out of nothing, but to extract energy from surrounding heat reservoirs in the earth, sea or air. For instance, you can imagine a steamship in whose boilers steam was gotten up, not by burning coal but by extracting heat from the surrounding water. In fact, if it were possible to force heat to flow away from cold toward greater heat, instead of the other way round, one could construct a system for pumping in sea-water, depriving it of its heat content, and disposing of the residue blocks of ice overboard. When a gallon of cold water freezes into ice, it gives off enough heat to raise another gallon of cold water almost to the boiling point. By pumping through several gallons of sea-water per minute, one could easily collect enough heat to run a good-sized engine. For all practical purposes, such a perpetual motion machine of the second type would be just as good as the kind designed to create energy out of nothing. With engines like this to do the work, everyone in the world could live as carefree an existence as a man with an unbeatable roulette system. Unfortunately they are equally impossible as they both violate the laws of probability in the same way.'

'I admit that trying to extract heat out of sea-water to raise steam in a ship's boilers is a crazy idea,' said Mr Tompkins. 'However, I fail to see any connexion between that problem and the laws of chance. Surely, you are not suggesting that dice and roulette wheels should be used as moving parts in these fuel-less machines. Or are you?'

'Of course not!' laughed the professor. 'At least I don't believe even the craziest perpetual motion inventor has made that suggestion yet. The point is that heat processes themselves are very similar in their nature to games of dice, and to hope that heat will flow from the colder body into the hotter one is like hoping that money will flow from the casino's bank into your pocket.'

'You mean that the bank is cold and my pocket hot?' asked Mr Tompkins, by now completely befuddled.

'In a way, yes,' answered the professor. 'If you hadn't missed my lecture last week, you would know that heat is nothing but the rapid irregular movement of innumerable particles, known as atoms and molecules, of which all material bodies are constituted. The more violent this molecular motion is, the warmer the body appears to us. As this molecular motion is quite irregular, it is subject to the laws of chance, and it is easy to show that the most probable state of a system made up of a large number of particles will correspond to a more or less uniform distribution among all of them of the total available energy. If one part of the material body is heated, that is if the molecules in this region begin to move faster, one would expect that, through a large number of accidental collisions, this excess energy would soon be distributed evenly among all the remaining particles. However, as the collisions are purely accidental, there is also the possibility that, merely by chance, a certain group of particles may collect the larger part of the available energy at the expense of the others. This spontaneous concentration of thermal energy in one particular part of the body would correspond to the flow of heat against the temperature gradient, and is not excluded in principle. However, if one tries to calculate the relative probability of such a spontaneous heat concentration occurring, one gets such small numerical values that the phenomenon can be labelled as practically impossible.'

'Oh, I see it now,' said Mr Tompkins. 'You mean that these perpetual motion machines of the second kind might work once in a while but that the chances of that happening are as slight as they are of throwing a seven a hundred times in a row in a dice game.'

'The odds are much smaller than that,' said the professor. 'In fact, the probabilities of gambling successfully against nature are so slight that it is difficult to find words to describe them. For instance, I can work out the chances of all the air in this room collecting spontaneously under the table, leaving an absolute vacuum everywhere else. The number of dice you would throw at one time would be equivalent to the number of air molecules in the room, so I must know how many there are. One cubic centimetre of air at atmospheric pressure, I remember, contains a number of molecules described by a figure of twenty digits, so the air molecules in the whole room must total a number with some twenty-seven digits. The space under the table is about one per cent of the volume of the room, and the chances of any given molecule being under the table and not somewhere else are, therefore, one in a hundred. So, to work out the chances of all of them being under the table at once, I must multiply one hundredth by one hundredth and so on, for each molecule in the room. My result will be a decimal beginning with fifty-four noughts.'

'Phew...!' sighed Mr Tompkins, 'I certainly wouldn't bet on those odds! But doesn't all this mean that deviations from equipartition are simply impossible?'

'Yes,' agreed the professor. 'You can take it as a fact that we won't suffocate because all the air is under the table, and for that matter that the liquid won't start boiling by itself in your high-ball glass. But if you consider much smaller areas, containing much smaller numbers of our dice-molecules, deviations from statistical distribution become much more probable. In this very room, for instance, air molecules habitually group themselves somewhat more densely at certain points, giving rise to minute inhomogeneities, called statistical fluctuations of density. When the sun's light passes through terrestrial atmosphere, such inhomogeneities cause the scattering of the blue rays of spectrum, and give to the sky its familiar colour. Were these fluctuations of density not present, the sky would always be quite black, and the stars would be clearly visible in full daylight. Also the slightly opalescent light liquids get when they are raised close to the boiling point is explained by these same fluctuations of density produced by the irregularity of molecular motion. But, on a large scale, such fluctuations are so extremely improbable that we would watch for billions of years without seeing one.'

'But there is still a chance of the unusual happening right now in this very room,' insisted Mr Tompkins. 'Isn't there?'

'Yes, of course there is, and it would be unreasonable to insist that a bowl of soup couldn't spill itself all over the table cloth because half of its molecules had accidentally received thermal velocities in the same direction.'

'Why that very thing happened only yesterday,' chimed in Maud, taking an interest now she had finished her magazine. 'The soup spilled and the maid said she hadn't even touched the table.'

The professor chuckled. 'In this particular case,' he said, 'I suspect the maid, rather than Maxwell's Demon, was to blame.'

'Maxwell's Demon?' repeated Mr Tompkins, surprised. 'I should think scientists would be the last people to get notions about demons and such.'

'Well, we don't take him very seriously,' said the professor. 'CLERK MAXWELL, the famous physicist, was responsible for introducing the notion of such a statistical demon simply as a figure of speech. He used this notion to illustrate discussions on the phenomena of heat. Maxwell's Demon is supposed to be rather a fast fellow, and capable of changing the direction of every single molecule in any way you prescribe. If there really were such a demon, heat could be made to flow against temperature, and the fundamental law of thermodynamics, the *principle of increasing entropy*, wouldn't be worth a nickel.'

'Entropy?' repeated Mr Tompkins. 'I've heard that word before. One of my colleagues once gave a party, and after a few drinks, some chemistry students he'd invited started singing—

> '*In*creases, *de*creases
> *De*creases, *in*creases
> What the hell do we care
> What entropy does?'

to the tune of "Ach du lieber Augustine". What is entropy anyway?'

'It's not difficult to explain. "Entropy" is simply a term used to describe the degree of disorder of molecular motion in any given physical body or system of bodies. The numerous irregular collisions between the molecules tend always to increase the entropy, as an absolute disorder is the most probable state of any statistical ensemble. However, if Maxwell's Demon could be put to work, he would soon put some order into the movement of the molecules the way a good sheep dog rounds up and steers a flock of sheep, and the entropy would begin to decrease. I should also tell you that according to the so-called H-theorem Ludwig Boltzmann introduced to science....'

Apparently forgetting he was talking to a man who knew practically nothing about physics and not to a class of advanced students, the professor rambled on, using such monstrous terms as 'generalized parameters' and 'quasi-ergodic systems', thinking he was making the fundamental laws of thermodynamics and their relation to Gibbs' form of statistical mechanics crystal clear. Mr Tompkins was used to his father-in-law talking over his head, so he sipped his Scotch and soda philosophically and tried to look intelligent. But all these highlights of statistical physics were definitely too much for Maud, curled up in her chair and struggling to keep her eyes open. To throw off her drowsiness she decided to go and see how dinner was getting along.

'Does madam desire something?' inquired a tall, elegantly dressed butler, bowing as she came into the dining room.

'No, just go on with your work,' she said, wondering why on earth he was there. It seemed particularly odd as they had never had a butler and certainly could not afford one. The man was tall and lean with an olive skin, long, pointed nose, and greenish eyes which seemed to burn with a strange, intense glow. Shivers ran up and down Maud's spine when she noticed the two symmetrical lumps half hidden by the black hair above his forehead.

'Either I'm dreaming,' she thought, 'or this is Mephistopheles himself, straight out of grand opera.'

'Did my husband hire you?' she asked aloud, just for something to say.

'Not exactly,' answered the strange butler, giving a last artistic touch to the dinner table. 'As a matter of fact, I came here of my own accord to show your distinguished father I am not the myth he believes me to be. Allow me to introduce myself. I am Maxwell's Demon.'

'Oh!' breathed Maud with relief, 'Then you probably aren't wicked, like other demons, and have no intention of hurting anybody.'

'Of course not,' said the Demon with a broad smile, 'but I like to play practical jokes and I'm about to play one on your father.'

'What are you going to do?' asked Maud, still not quite reassured.

'Just show him that, if I choose, the law of increasing entropy can be broken. And to convince you it can be done, I would appreciate the honour of your company. It is not at all dangerous, I assure you.'

At these words, Maud felt the strong grip of the Demon's hand on her elbow, and everything around her suddenly went crazy. All the familiar objects in her dining room began to grow with

'Is this what hell looks like?'

terrific speed, and she got a last glimpse of the back of a chair covering the whole horizon. When things finally quieted down, she found herself floating in the air supported by her companion. Foggy-looking spheres, about the size of tennis balls, were whizzing by in all directions, but Maxwell's Demon cleverly kept them from colliding with any of the dangerous looking things. Looking down, Maud saw what looked like a fishing boat, heaped to the gunwales with quivering, glistening fish. They were not fish, however, but a countless number of foggy balls, very like those flying past them in the air. The Demon led her closer until she seemed surrounded by a sea of coarse gruel which was moving and working in a patternless way. Balls were boiling to the surface and others seemed to be sucked down. Occasionally one would come to the surface with such speed it would tear off into space, or one of the balls flying through the air would dive into the gruel and disappear under thousands of other balls. Looking at the gruel more closely, Maud discovered that the balls were really of two different kinds. If most looked like tennis balls, the larger and more elongated ones were shaped more like American footballs. All of them were semi-transparent and seemed to have a complicated internal structure which Maud could not make out.

'Where are we?' gasped Maud. 'Is this what hell looks like?'

'No,' smiled the Demon, 'Nothing as fantastic as that. We are simply taking a close look at a very small portion of the liquid surface of the highball which is succeeding in keeping your husband awake while your father expounds quasi-ergodic systems. All these balls are molecules. The smaller round ones are water molecules and the larger, longer ones are molecules of alcohol. If you care to work out the proportion between their number, you can find out just how strong a drink your husband poured himself.'

'*Very* interesting,' said Maud, as sternly as she dared. 'But what are those things over there that look like a couple of whales playing in the water. They couldn't be atomic whales, or could they?'

The demon looked where Maud pointed. 'No, they are hardly whales,' he said. 'As a matter of fact, they are a couple of very fine fragments of burned barley, the ingredient which gives whisky its particular flavour and colour. Each fragment is made up of millions and millions of complex organic molecules and is comparatively large and heavy. You see them bouncing around because of the action of impacts they receive from the water and alcohol molecules animated by thermal motion. It was the study of such intermediate-sized particles, small enough to be influenced by molecular motion but still large enough to be seen through a strong microscope, which gave scientists their first direct proof of the kinetic theory of heat. By measuring the intensity of the

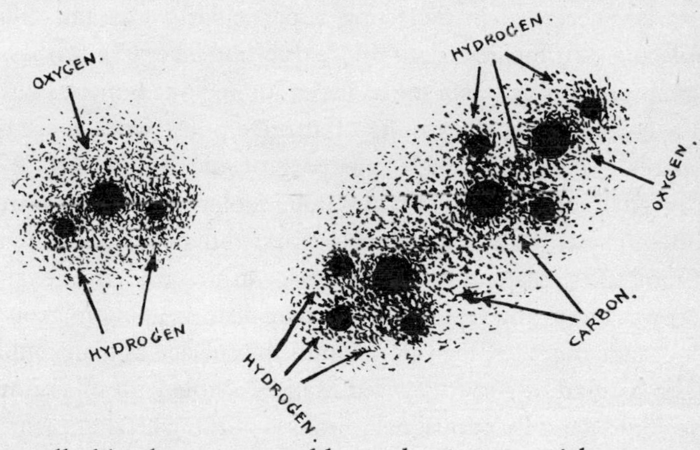

tarantella-like dance executed by such minute particles suspended in liquids, their Brownian motion as it is usually called, physicists were able to get direct information on the energy of molecular motion.'

Again the Demon guided her through the air until they came to an enormous wall made of numberless water molecules fitted neatly and closely together like bricks.

'How very impressive!' cried Maud. 'That's just the background I've been looking for for a portrait I'm painting. What is this beautiful building, anyway?'

'Why, this is part of an ice crystal, one of many in the ice cube in your husband's glass,' said the Demon. 'And now, if you will excuse me, it is time for me to start my practical joke on the old, self-assured professor.'

So saying, Maxwell's Demon left Maud perched on the edge of the ice crystal, like an unhappy mountain climber, and set about his work. Armed with an instrument like a tennis racquet, he was swatting the molecules around him. Darting here and there, he was always in time to swat any stubborn molecule which persisted in going in the wrong direction. In spite of the apparent danger of her position, Maud could not help admiring his wonderful speed and accuracy, and found herself cheering with excitement whenever he succeeded in deflecting a particularly fast and difficult molecule. Compared with the exhibition she was witnessing, champion tennis players she had seen looked like hopeless duffers. In a few minutes, the results of the Demon's work were quite apparent. Now, although one part of the liquid surface was covered by very slowly moving, quiet molecules, the part directly under her feet was more furiously agitated than ever. The number of molecules escaping from the surface in the process of evaporation was increasing rapidly. They were now escaping in groups of thousands together, tearing through the surface as giant bubbles. Then a cloud of steam covered Maud's whole field of vision and she could get only occasional glimpses of the whizzing racquet or the tail of the Demon's dress suit among the masses of maddened molecules. Finally the molecules in her ice crystal perch gave way and she fell into the heavy clouds of vapour beneath....

When the clouds cleared, Maud found herself sitting in the same chair she was sitting in before she went into the dining room.

'Holy entropy!' her father shouted, staring bewildered at Mr Tompkins' highball. 'It's boiling!'

The liquid in the glass was covered with violently bursting bubbles, and a thin cloud of steam was rising slowly toward the ceiling. It was particularly odd, however, that the drink was boiling only in a comparatively small area around the ice cube. The rest of the drink was still quite cold.

'Think of it!' went on the professor in an awed, trembling voice. 'Here I was telling you about statistical fluctuations in the law of entropy when we actually see one! By some incredible chance, possibly for the first time since the earth began, the faster molecules have all grouped themselves accidentally on one part of the surface of the water and the water has begun to boil by itself!

'Holy entropy! It's boiling!'

In the billions of years to come, we will still, probably, be the only people who ever had the chance to observe this extraordinary phenomenon.' He watched the drink, which was now slowly cooling down. 'What a stroke of luck!' he breathed happily.

Maud smiled but said nothing. She did not care to argue with her father, but this time she felt sure she knew better than he.

10

The Gay Tribe of Electrons

A few days later, while finishing his dinner, Mr Tompkins remembered that it was the night of the professor's lecture on the structure of the atom, which he had promised to attend. But he was so fed up with his father-in-law's interminable expositions that he decided to forget the lecture and spend a comfortable evening at home. However, just as he was getting settled with his book, Maud cut off this avenue of escape by looking at the clock and remarking, gently but firmly, that it was almost time for him to leave. So, half an hour later, he found himself on a hard wooden bench in the university auditorium together with a crowd of eager young students.

'Ladies and gentlemen,' began the professor, looking at them gravely over his spectacles, 'In my last lecture I promised to give you more details concerning the internal structure of the atom, and to explain how the peculiar features of this structure account for its physical and chemical properties. You know, of course, that atoms are no longer considered as elementary indivisible constituent parts of matter, and that this role has passed now to much smaller particles such as electrons, protons, etc.

'The idea of elementary constituent particles of matter, representing the last possible step in divisibility of material bodies, dates back to the ancient Greek philosopher DEMOCRITUS who lived in the fourth century B.C. Meditating about the hidden nature of things, Democritus came to the problem of the structure of matter and was faced with the question whether or not it can exist in infinitely small portions. Since it was not customary at this epoch to solve any problem by any other method than that of pure thinking, and since, in any case, the question was at that time beyond any possible attack by experimental methods,

Democritus searched for the correct answer in the depths of his own mind. On the basis of some obscure philosophical considerations he finally came to the conclusion that it is "unthinkable" that matter could be divided into smaller and smaller parts without any limit, and that one must assume the existence of "the smallest particles which cannot be divided any more". He called such particles "atoms", which, as you probably know, means "indivisibles" in Greek.

'I do not want to minimize the great contribution of Demo critus to the progress of natural science, but it is worth keeping in mind that besides Democritus and his followers, there was undoubtedly another school of Greek philosophy the adherents of which maintained that the process of divisibility of matter *could* be carried beyond any limit. Thus, independent of the character of the answer which had to be given in the future by exact science, the philosophy of ancient Greece was well secured with an honourable place in the history of physics. At the time of Democritus, and for centuries later, the existence of such indivisible portions of matter represented a purely philosophical hypothesis, and it was only in the nineteenth century that scientists decided that they had finally found these indivisible building-stones of matter which were foretold by the old Greek philosopher more than two thousand years ago.

'In fact, in the year 1808 an English chemist, JOHN DALTON, showed that the relative proportions....'

Almost from the beginning of the lecture Mr Tompkins had felt an irresistible urge to close his eyes and doze through the rest of the lecture, and it was only the academic hardness of the bench that kept him from doing so. However, Dalton's ideas concerning the law of 'relative proportions' proved the last straw, and the hushed auditorium was soon permeated by a gentle wheeze coming from the corner where Mr Tompkins was sitting.

When Mr Tompkins dropped off to sleep, the discomfort of the uncompromising bench seemed to melt into the pleasant sensation of floating on air, and opening his eyes he was surprised to find himself dashing through space at what he considered a pretty reckless speed. Looking around he saw that he was not alone on this fantastic trip. Near him a number of vague, misty forms were swooping around a large heavy-looking object in the middle of the crowd. These strange beings were travelling in pairs, gaily

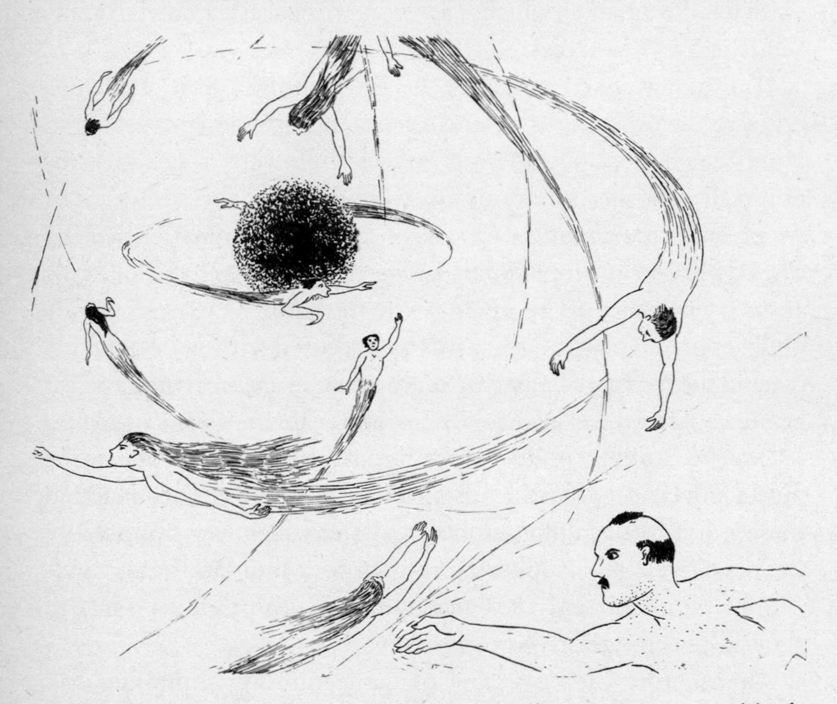

chasing each other along circular and elliptic tracks. Suddenly Mr Tompkins felt very lonely because he realized that he was the only one of the whole group who had no playmate.

'Why didn't I bring Maud along with me?' Mr Tompkins wondered gloomily. 'We could have had a wonderful time with this happy-go-lucky crowd.' The track he was moving along was outside all the others, and while he wanted very much to join the party, the uncomfortable feeling of being odd man out kept him from doing so. However, when one of the electrons (for by now

Mr Tompkins realized he had miraculously joined the electronic community of an atom) was passing close by on its elongated track, he decided to complain about the situation.

'Why haven't *I* got anyone to play with?' he shouted across.

'Because this is an odd atom, and you are the valency elec-tr-o-o-on...,' called the electron as it turned and plunged back into the dancing crowd.

'Valency electrons live alone or find companions in other atoms,' squeaked the high pitched soprano of another electron rushing past him.

'If you want a partner fair,
Jump into chlorine and find one there,'

chanted another mockingly.

'I see you are quite new here, my son, and very lonely,' said a friendly voice above him, and raising his eyes Mr Tompkins saw the stout figure of a monk clothed in a brown tunic.

'I am Father Paulini,' went on the monk, moving along the track with Mr Tompkins, 'and it is my mission in life to keep watch over the morals and social life of electrons in atoms and elsewhere. It is my duty to keep these playful electrons properly distributed among the different quantum cells of the beautiful atomic structures erected by our great architect Niels Bohr. To keep order and to preserve the proprieties, I never permit more than two electrons to follow the same track; a *ménage à trois* always gives a lot of trouble, you know. Thus electrons are always grouped in pairs of opposite "spin" and no intruder is permitted if the cell is already occupied by a couple. It is a good rule, and I may add that not a single electron has yet broken my commandment.'

'Maybe it *is* a good rule,' objected Mr Tompkins, 'but it is rather inconvenient for me at the moment.'

'I see it is,' smiled the monk, 'but it is just your bad luck, being a valency electron in an odd atom. The sodium atom to which you belong is entitled by the electric charge of its nucleus (that big dark mass you see in the centre) to hold eleven electrons together.

Well, unfortunately, for you eleven is an odd number, hardly an unusual circumstance when you consider that exactly one half of all numbers are odd, and only the other half even. Thus, as the latecomer you will have to be alone for a while at least.'

'You mean there is a chance that I can get in later?' asked Mr Tompkins eagerly. 'Kicking one of the oldtimers out, for example?'

'It isn't exactly done,' said the monk wagging a plump finger at him, 'but, of course, there is always a chance that some of the inner circle members will be thrown out by an external disturbance, leaving an empty place. However, I wouldn't count on it much, if I were you.'

'They told me I'd be better off if I moved into chlorine,' said Mr Tompkins, discouraged by Father Paulini's words. 'Can you tell me how to do that?'

'Young man, young man!' exclaimed the monk sorrowfully, 'why are you so insistent on finding company? Why can't you appreciate solitude and this Heaven-sent opportunity to contemplate your soul in peace? Why must even electrons lean always to the worldly life? However, if you insist on companionship, I will help you to get your wish. If you look where I'm pointing, you will see a chlorine atom approaching us, and even at this distance you can see an unoccupied spot where you would most certainly be welcomed. The empty spot is in the outer group of electrons, the so-called "M-shell", which is supposed to be made up of eight electrons grouped in four pairs. But, as you see, there are four electrons spinning in one direction and only three in the other, with one place vacant. The inner shells, known as "K" and "L", are completely filled up, and the atom will be glad to get you and have its outer shell complete. When the two atoms get close together, just jump over, as valency electrons usually do. And may peace be with you, my son!' With these words the impressive figure of the electron priest suddenly faded into thin air.

Feeling considerably more cheerful, Mr Tompkins gathered his strength for a neckbreaking jump into the orbit of the passing chlorine atom. To his surprise he leapt over with an easy grace and found himself in the congenial surroundings of the members of the chlorine M-shell.

'It was delightful of you to join us!' called his new partner of opposite spin, gliding gracefully along the track. 'Now no one can say that our community is not complete. Now we shall all have fun together!'

Mr Tompkins agreed that it really was fun—lots of fun—but

one little worry kept stealing into his mind. 'How am I going to explain this to Maud when I see her again?' he thought rather guiltily, but not for long. 'Surely she won't mind,' he decided. 'After all, these are only electrons.'

'Why doesn't that atom you've left go away now?' asked his companion with a pout. 'Does it still hope to get you back?'

And, as a matter of fact, the sodium atom, with its valency electron gone, *was* sticking closely to the chlorine one as if in the hope that Mr Tompkins would change his mind and jump back to his lonely track.

'Well how do you like that!' said Mr Tompkins angrily, frowning at the atom which had first received him so coldly. 'There's a dog in the manger for you!'

'Oh, they always do that,' said a more experienced member of the M-shell. 'I understand it is not so much the electronic community of the sodium atom which wants you back as the sodium nucleus itself. There is almost always some disagreement between the central nucleus and its electronic escort: the nucleus wants as many electrons around it as it can possibly hold with its electric charge, whereas the electrons themselves prefer to be only enough in number to make the shells complete. There are only a few atomic species, the so-called *rare gases*, or *noble gases* as the German chemists call them, in which the desire of the ruling nucleus and the subordinate electrons are in full harmony. Such atoms as helium, neon and argon, for example, are quite satisfied with themselves and neither expel their number nor invite new ones. They are chemically inert, and keep away from all other atoms. But in all other atoms electronic communities are always ready to change their membership. In the sodium atom, which was your former home, the nucleus is entitled by its electric charge to one more electron than is necessary for harmony in the shells. On the other hand, in our atom the normal contingent of electrons is not enough for complete harmony, and thus we welcome your arrival, in spite of the fact that your presence overloads our nucleus. But as long as you stay here, our atom is not neutral any more, and has an extra electric charge. Thus the sodium atom which you left stands by, held by the force of electric attraction. I once heard our great priest, Father Paulini, say that such atomic communities, with extra electrons or electrons missing, are called negative and positive "ions". He also uses the word "molecule" for groups of two or more atoms bound together by electric force. This particular combination of sodium and chlorine atoms he calls a molecule of "table salt", whatever that may be.'

'Do you mean to tell me you don't know what table salt is?' said Mr Tompkins, forgetting to whom he was talking. 'Why that's what you put on your scrambled eggs at breakfast.'

'What are "scram bulldeggs" and what is "break-fust"?' asked the intrigued electron. Mr Tompkins sputtered and then realized the futility of trying to explain to his companions even the simplest details of the lives of human beings. 'That's why I don't get more out of their talk about valency and complete shells,' he told himself, deciding to enjoy his visit to this fantastic world without worrying about understanding it. But it was not so easy to get away from the talkative electron, who evidently had a great desire to pass on all the knowledge collected during a long electronic life.

'You must not think,' he continued, 'that the binding of atoms into molecules is always accomplished by one valency electron alone. There are atoms, like oxygen for example, which need two more electrons to complete their shells, and there are also atoms which need three electrons and even more. On the other hand, in some atoms the nucleus holds two or more extra—or valency—electrons. When such atoms meet, there is quite a lot of jumping over and binding to do, as a result of which quite complex molecules, often consisting of thousands of atoms, are formed. There are also the so-called "homopolar" molecules, that is molecules made up of two identical atoms, but that is a very unpleasant situation.'

'Unpleasant, why?' asked Mr Tompkins, getting interested again.

'Too much work,' commented the electron, 'to keep them together. Some time ago I happened to get that job and I didn't have a moment to myself all the while I stayed there. Why, it isn't at all the way it is here where the valency electron just enjoys himself and lets the electrically hungry and deserted atom stand by. No sir! In order to keep the two identical atoms together, he has to jump to and fro, from one to the other and back again. My word! One feels like a ping pong ball.'

Mr Tompkins was rather surprised to hear the electron, which did not know what scrambled eggs were, speak so glibly of ping pong, but he let it pass.

'I'll never take on that job again!' grumbled the lazy electron, overwhelmed by a wave of unpleasant memories. 'I am quite comfortable where I am now.'

'Wait!' he exclaimed suddenly. 'I think I see a still better place for me to go. So lo-o-o-ong!' And with a giant leap he rushed toward the interior of the atom.

Looking in the direction in which his interlocutor had gone, Mr Tompkins understood what had happened. It seems that one of the electrons of the inner circle was thrown clear of the atom by some foreign high-speed electron which had unexpectedly penetrated into their system, and a cosy place in the 'K' shell was now wide open. Chiding himself for missing this opportunity to join the inner circle, Mr Tompkins now watched with great interest the course of the electron he had just been talking to. Deeper and deeper into the atomic interior this happy electron sped, and bright rays of light accompanied his triumphant flight. Only when it finally reached the internal orbit did this almost unbearable radiation finally stop.

'What was that?' asked Mr Tompkins, his eyes aching from the sight of this unexpected phenomenon. 'Why all this brilliance?'

'Oh that's just the X-ray emission connected with the transition,' explained his orbit companion, smiling at his embarrassment. 'Whenever one of us succeeds in getting deeper into the interior of the atom, the surplus energy must be emitted in the form of radiation. This lucky fellow made quite a big jump and let loose a lot of energy. More often we have to be satisfied with smaller jumps, here in the atomic suburbs, and then our radiation is called "visible light"—at least that is what Father Paulini calls it.'

'But this X-light, or whatever you call it, is also visible,' protested Mr Tompkins. 'I should call your terminology rather misleading.'

'Well, we are electrons and are susceptible to any kind of radiation. But Father Paulini tells us that there exist gigantic creatures, "Human Beings", he calls them, who can see light only when it falls within a narrow energy-interval, or wave length-interval as he puts it. He told us once that it took a great man, Roentgen I think his name was, to discover these X-rays and that now they are largely used in something called "medicine".'

'Oh yes. I know quite a lot about that,' said Mr Tompkins, feeling proud that now *he* could show off his knowledge. 'Want me to tell you more about it?'

'No thanks,' said the electron yawning. 'I really don't care. Can't you be happy without talking? Try to catch me!'

For a long time Mr Tompkins went on enjoying the pleasant sensation of diving through space with the other electrons in a kind of glorified trapeze act. Then, all of a sudden, he felt his hair stand on end, an experience he had felt once before during a thunder storm in the mountains. It was clear that a strong electric disturbance was approaching their atom, breaking the harmony of the electronic motion, and forcing the electrons to deviate seriously from their normal tracks. From the point of view of a human physicist, it was only a wave of ultraviolet light passing through the spot where this particular atom happened to be, but to the tiny electrons it was a terrific electric storm.

'Hold on tight!' yelled one of his companions, 'or you will be thrown out by photo-effect forces!' But it was already too late. Mr Tompkins was snatched away from his companions and hurled into space at a terrifying speed, as neatly as if he had been seized by a pair of powerful fingers. Breathlessly he hurtled further and further through space, tearing past all kinds of different atoms so fast he could hardly distinguish the separate electrons. Suddenly a large atom loomed up right in front of him and he knew that a collision was unavoidable.

'Pardon me, but I am photo-effected and cannot...,' began Mr Tompkins politely, but the rest of the sentence was lost in an ear-splitting crash as he ran head on into one of the outer electrons. The two of them tumbled head over heels off into space. However, Mr Tompkins had lost most of his speed in the collision and was now able to study his new surroundings somewhat more closely. The atoms which towered around him were much larger than any he had seen before, and he could count as many as twenty-nine electrons in each of them. If he had known his physics better he would have recognized them as atoms of copper, but at these close quarters the group as a whole did not look like copper at all. Also they were spaced rather close to one another forming a regular pattern which extended as far as he could see. But what surprised Mr Tompkins most was the fact that these atoms did not seem to be very particular about holding on to their quota of electrons, particularly their outer electrons. In fact the outer orbits were mostly empty, and crowds of unattached electrons were drifting lazily through space, stopping from time to time but never for very long, on the outskirts of one atom or another. Rather tired after his breakneck flight through space, Mr Tompkins tried at first to get a little rest on a steady orbit of one of the copper atoms. However he was soon infected with the prevailing vagabondish feeling of the crowd, and he joined the rest of the electrons in their nowhere-in-particular motion.

'Things are not very well organized here,' he commented to

himself, 'and there are too many electrons not tending to their business. I think Father Paulini should do something about it.'

'Why should I?' said the familiar voice of the monk who had suddenly materialized from nowhere. 'These electrons are not disobeying my commandments, and besides they are doing a very useful job indeed. You may be interested to know that if all atoms cared as much about holding their electrons as some of them do, there would be no such thing as electric conductivity. Why you wouldn't even be able to have an electric bell in your house, to say nothing of a light or a telephone.'

'Oh, you mean these electrons carry electric current?' asked Mr Tompkins, grasping at the hope that the conversation was turning to a subject more or less familiar to him. 'But I don't see that they are moving in any particular direction.'

'First of all, my lad,' said the monk severely, 'do not use the word "they", use "we". You seem to forget that you are an electron yourself and that the moment someone presses the button to which this copper wire is attached, electric tension will cause you, as well as all the other conductivity electrons, to rush along to call the maid or do whatever else is needed.'

'But I don't want to!' said Mr Tompkins firmly, a note of temper in his voice. 'As a matter of fact I am quite tired of being an electron and I don't think it's so much fun any more. What a life, to have to carry out all these electronic duties for ever and ever!'

'Not necessarily forever,' countered Father Paulini, who definitely did not like back-talk on the part of plain electrons. 'There is always the chance that you will be annihilated and cease to exist.'

'B-b-be annihilated?' repeated Mr Tompkins feeling cold shivers running up and down his spine. 'But I always thought electrons were eternal.'

'That is what physicists used to believe until comparatively recent times,' agreed Father Paulini, amused at the effect produced by his words, 'but it isn't exactly correct. Electrons can be born, and die, as well as human beings. There isn't, of course, such a thing as dying of old age; death comes only through collisions.'

'Well, I had a collision only a short while ago, and a pretty bad one too,' said Mr Tompkins recovering a little confidence. 'And if that one didn't put me out of action, I can't imagine one that would.'

'It isn't a question of how forcibly you collide,' Father Paulini corrected him, 'but of who the other fellow is. In your recent collision you probably ran into another negative electron, very similar to yourself, and there is not the slightest danger in such an encounter. In fact, you could butt into each other like a couple of rams for years and no harm could be done. But there is another breed of electron, the positive ones, which have been discovered only comparatively recently by the physicists. These positive electrons, or positrons, look exactly the way you do, the only difference being that their electric charge is positive instead of negative. When you see such a fellow approaching, you think it is just another innocent member of your tribe and go ahead to greet him. But then you suddenly find that, instead of pushing you away slightly to avoid a collision, as any normal electron would, he pulls you right in. And then it is too late to do anything.'

'How terrible!' exclaimed Mr Tompkins. 'And how many poor ordinary electrons can one positron eat up?'

'Fortunately only one, since in destroying a negative electron the positron also destroys itself. One could describe them as members of a suicide club, looking for partners in mutual annihiltion. They do not harm one another, but as soon as a negative electron comes their way, it hasn't much chance of surviving.'

'Lucky I haven't run into one of these monsters yet,' said Mr Tompkins much impressed by this description. 'I hope they are not very numerous. Are they?'

'No, they're not. And for the simple reason that they are always looking for trouble and so vanish very soon after they are born. If you wait a minute, I shall probably be able to show you one.'

'Yes, here we are,' continued Father Paulini after a short silence. 'If you look carefully at that heavy nucleus over there, you will see one of these positrons being born.'

The atom at which the monk was pointing was evidently undergoing a strong electromagnetic disturbance owing to some vigorous radiation falling on it from outside. It was a much more violent disturbance than the one which threw Mr Tompkins out of his chlorine atom, and the family of atomic electrons surrounding the nucleus was being dispersed and blown away like dry leaves in a hurricane.

'Look closely at the nucleus,' said Father Paulini, and concentrating his attention Mr Tompkins saw a most unusual phenomenon taking place in the depths of the destroyed atom. Very close to the nucleus, inside the inner electronic shell, two vague shadows were gradually taking shape, and a second later Mr Tompkins saw two glittering brand new electrons rushing at great speed away from their birthplace.

'But I see two of them,' said Mr Tompkins, fascinated by the sight.

'That is right,' agreed Father Paulini. 'Electrons are always born in pairs, otherwise it would contradict the law of conservation of electric charge. One of these two particles, born under the action of a strong gamma ray on the nucleus, is an ordinary negative electron, whereas the other is a positron—the murderer. He is off now to find a victim.'

'Well, if the birth of each positron destined to destroy an electron is accompanied by the birth of still another plain electron, then things aren't so bad,' commented Mr Tompkins thoughtfully. 'At least it doesn't lead to the extinction of the electronic tribe, and I. . . .'

'Look out!' interrupted the monk shoving him aside while the newborn positron whistled by, just an inch away. 'You can never be too careful when these murderous particles are around. But I think I'm spending too much time talking to you and I have other business to attend to. I must look for my pet "neutrino"....' And the monk disappeared without letting Mr Tompkins know what this 'neutrino' was and whether or not it was also to be feared. Thus deserted, Mr Tompkins felt even more lonely than before and, when one or another fellow electron approached him on his journey through space, he even nursed a secret desperate hope that under each innocent exterior might be

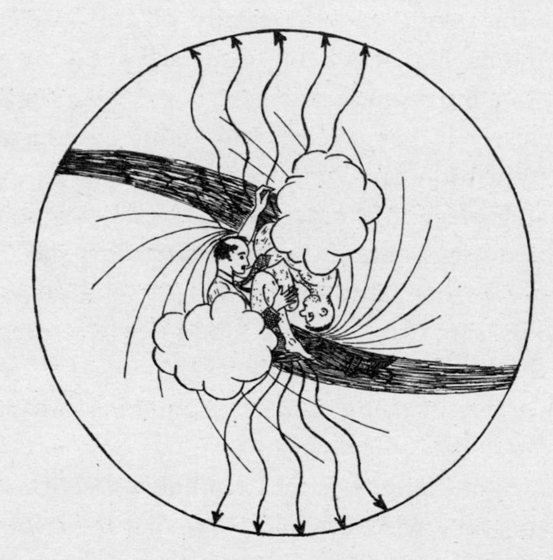

hidden the heart of a murderer. For a long time, centuries it seemed to him, his fears and hopes were not justified, and he unwillingly bore the dull duties of a conductivity electron.

Then suddenly it happened, and at a moment when he expected it least. Feeling a strong need to talk to somebody, even to a stupid conductivity electron, he approached a particle which was slowly moving by and was evidently a newcomer to this bit of copper wire. Even at a distance, however, he noticed that he had made a bad choice and that an irresistible force of attraction was pulling him along, permitting no retreat. For a second he tried to struggle and tear himself away, but the distance between them was rapidly getting smaller and smaller and it seemed to Mr Tompkins that he saw a fiendish grin on the face of his captor.

'Let me go! Let me go!' shouted Mr Tompkins at the top of his voice, struggling with his arms and kicking his legs. 'I don't want to be annihilated; I'll conduct electric current for the rest of eternity!' But it was all in vain, and the surrounding space was suddenly illuminated by a blinding flash of intensive radiation.

'Well, I am no more,' thought Mr Tompkins, 'but how is it I can still think? Has my body only been annihilated, and my soul gone to a quantum heaven?' Then he felt a new force, more gentle this time, shaking him firmly and resolutely, and opening his eyes he recognized the university janitor.

'I'm sorry, Sir,' he said, 'but the lecture was over some time ago and we gotta close the hall up now.' Mr Tompkins stifled a yawn and looked sheepish.

'Good night, Sir,' said the janitor with a sympathetic smile.

$10\frac{1}{2}$

A Part of the Previous Lecture which Mr Tompkins slept through

In fact, in the year 1808, an English chemist JOHN DALTON showed that the relative proportions of various chemical elements which are needed to form more complicated chemical compounds can always be expressed by the ratio of integral numbers, and he interpreted this empirical law as due to the fact that all compound substances are built up from a varying number of particles representing simple chemical elements. The failure of medieval alchemy to turn one chemical element into another supplied a proof of apparent indivisibility of these particles, and without much hesitation they were christened by the old Greek name: 'atoms'. Once given, the name stuck, and although we know now that these 'Dalton's atoms' are not at all indivisible, and are, in fact, formed by a large number of still smaller particles, we close our eyes to the philological inconsistency of their name.

Thus the entities called 'atoms' by modern physics are not at all the elementary and indivisible constituent units of matter imagined by Democritus, and the term 'atom' would actually be more correct if it were applied to such much smaller particles as electrons and protons, from which 'Dalton's atoms' are built. But such a change of names would cause too much confusion, and nobody in physics cares much about philological consistency anyway! Thus we retain the old name of 'atoms' in Dalton's sense, and refer to electrons, protons, etc. as 'elementary particles'.

This name indicates, of course, that we believe at present that these smaller particles are *really* elementary and indivisible in Democritus' sense of the word, and you may ask me whether history will not repeat itself, and whether in the further progress of science, the elementary particles of modern physics will not be proved to be quite complex. My answer is that, although there is no absolute guarantee that this will not happen, there are very good reasons to believe that this time we are completely right. In fact, there are ninety-two different kinds of atoms (corresponding to ninety-two different chemical elements) and each kind of atom possesses rather complicated characteristic properties; a situation which in itself invites some simplification along the line of reducing such a complicated picture to a more elementary one. On the other hand, physics of today recognizes only a few different kinds of elementary particles: *electrons* (positive and negative light particles), *nucleons* (charged or neutral heavy particles, also known as *protons and neutrons*), and possibly the so-called *neutrinos* the nature of which has not been completely clarified.

The properties of these elementary particles are extremely simple, and very little simplification could be gained by further reduction; besides, as you will understand, you must always have several elementary notions to play with if you want to build up something more complicated, and two or three elementary notions are not too many. Thus, in my opinion, it is quite safe to bet your last dollar that the elementary particles of modern physics will live up to their name.

Now we can turn to the question concerning the way in which Dalton's atoms are built up from the elementary particles. The first correct answer to this question was given in 1911 by the celebrated British physicist ERNEST RUTHERFORD (later Lord Rutherford of Nelson) who was studying atomic structure by bombarding various atoms with fast-moving minute projectiles, known as *alpha-particles*, which are emitted in the process of disintegration of radioactive elements. Observing the deflection (scattering) of these projectiles after passage through a piece of matter, Rutherford came to the conclusion that all atoms must possess a very dense positively charged central core (atomic nucleus) surrounded by a rather rarefied cloud of negative electric charge (atomic atmosphere). We know today that the atomic nucleus is made up of a certain number of *protons* and *neutrons*, known under the collective name of '*nucleons*', held tightly together by strong cohesive forces, and that atomic atmosphere consists of varying numbers of negative electrons swarming around under the action of electrostatic attraction of the nuclear positive charge. The number of electrons forming the atomic atmosphere determines all the physical and chemical properties of a given atom, and varies along the natural sequence of chemical elements from one (for hydrogen) up to ninety-two (for the heaviest known element: Uranium).

In spite of the apparent simplicity of Rutherford's atomic model, its detailed understanding turned out to be anything but simple. In fact, according to the best belief of classical physics, negatively charged electrons rotating around an atomic nucleus are bound to lose their energy of motion through the process of radiation (light-emission), and it has been calculated that, owing to these steady energy losses, all electrons forming atomic atmosphere should collapse on the nucleus within a negligible fraction of a second. This seemingly sound conclusion of classical theory stands, however, in sharp contradiction with the empirical fact that atomic atmospheres are, on the contrary, quite stable, and that, instead of collapsing on the nucleus, atomic electrons continue their swarming motion around the central body for an indefinite period of time. Thus we see that a very deep-rooted conflict arises between the basic ideas of classical mechanics, and the empirical data pertaining to the mechanical behaviour of a tiny constituent part in the world of atoms. This fact brought the famous Danish physicist Niels Bohr to the realization that classical mechanics, which claimed for centuries a privileged and secure position in the system of natural sciences, should be from now on considered as a restricted theory, applicable to the macroscopic world of our everyday experience, but failing badly in its application to the much more delicate types of motion taking place within various atoms. As the tentative foundation for the new generalized mechanics which would be applicable also to the motion of the tiny moving parts of atomic mechanism, Bohr proposed to assume that *from all the infinite variety of types of motion considered in classical theory, only a few specially selected types can actually take place in nature.* These permitted types of motion, or trajectories, are to be selected according to certain mathematical conditions, known as the *quantum conditions* of the Bohr theory. I am not going to enter here into a detailed discussion of these quantum conditions, but will mention only that they have been chosen in such a way, that all the restrictions imposed by them become of no practical importance in all cases where the mass of the moving particles is much larger than the masses we encounter in atomic structure. Thus, being applied to macroscopic bodies, the new *micro-mechanics* gives exactly the same results as the old classical theory (*principle of correspondence*) and it is only in the case of tiny atomic mechanisms that the disagreement between the two theories becomes of essential value. Without going deeper into the details, I will satisfy your curiosity concerning the structure of the atom from the point of view of Bohr's theory, by showing the diagram of Bohr's quantum orbits in an atom. (First plate, please!) You see here [p. 132], on a largely magnified scale of course, the system of circular and elliptical orbits, which represent the only types of motion 'permitted' for the electrons forming atomic atmosphere by Bohr's quantum conditions. Whereas classical mechanics would allow the electron to move at *any* distance from the nucleus and puts *no restriction* on the eccentricity (i.e. elongation) of its orbit, the selected orbits of Bohr's theory form a discrete set with all their characteristic dimensions sharply defined. Numbers and letters standing near each orbit indicate the name of any given orbit in the general classifications; you may notice, for example, that larger numbers correspond to the orbits of larger diameters.

Although Bohr's theory of atomic structure turned out to be

extremely fruitful in the explanation of various properties of atoms and molecules, the fundamental notion of discrete quantum orbits remained rather unclear, and the deeper we tried to go into the analysis of this unusual restriction of classical theory, the less clear was the entire picture.

Thus we obtain the original Bohr–Sommerfeld scheme for the permitted quantum orbits of an electron in a hydrogen atom

It finally became clear that the disadvantage of Bohr's theory lay in the fact that, *instead of changing* classical mechanics in some fundamental way, it *was simply restricting* the results of this system by additional conditions which were in principle foreign to the whole structure of classical theory. The correct solution of the entire problem came only thirteen years later, in the form of so-called 'wave-mechanics', which has modified the entire basis of classical mechanics in accordance with the new quantum-principle. And, in spite of the fact that at first sight the system of wave-mechanics may seem still crazier than Bohr's old theory, this new micro-mechanics represents one of the most consistent and accepted parts of the theoretical physics of today. Since the fundamental principle of the new mechanics, and in particular the notions of 'indeterminacy' and 'spreading out trajectories', have been already discussed by me in one of my previous lectures, I will refer you to your memory or your notes, and will return to

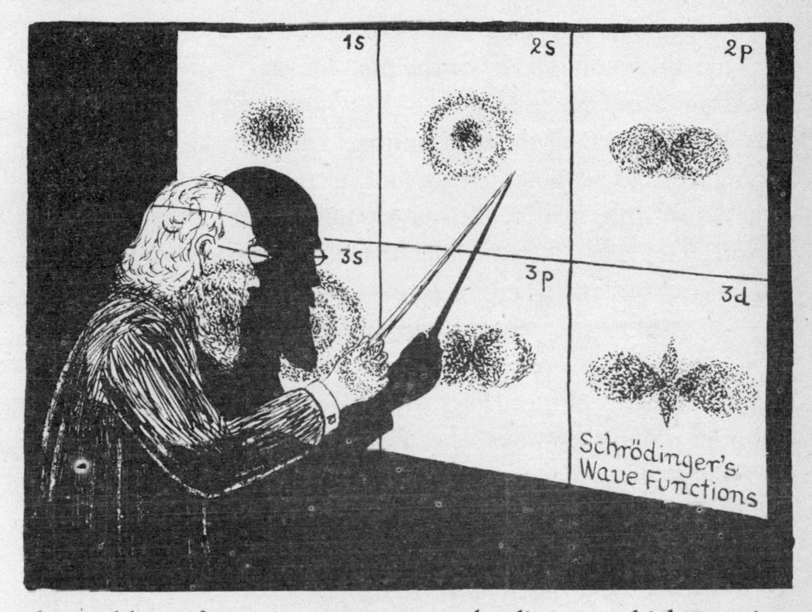

the problem of atomic structure. In the diagram which I project now (second plate, please!) you see the way in which the motion of atomic electrons is visualized by wave-mechanical theory from the point of view of 'spreading out trajectories'. This picture represents the same types of motion as those represented classically in the previous diagram (apart from the fact that for technical reasons each type of motion is now drawn separately), but instead of the sharp-lined trajectories of Bohr's theory, we have now diffuse patterns consistent with the fundamental *uncertainty principle*. The notations of different states of motion is the same as on the previous diagram, and, comparing the two, you will notice, if you will stretch your imagination slightly, that our cloudy form repeats rather faithfully the general features of the old Bohr's orbits.

These diagrams show you quite clearly what happens to the good old-fashioned trajectories of classical mechanics when the quantum is at play, and although a layman might think it a fantastic dream, scientists working in the microcosmos of atoms do not experience any difficulties in accepting this picture.

After this short survey of the possible states of motion in the electronic atmosphere of an atom, we now come to an important problem concerning the distribution of various atomic electrons among various possible states of motion. Here again we encounter a new principle, a principle quite unfamiliar in the macroscopic world. This principle was first formulated by my young friend WOLFGANG PAULI, and states that *in the community of electrons of a given atom no two particles may simultaneously possess the same type of motion.* This restriction would be of no great importance if, as it is in classical mechanics, there were an infinity of possible motions. Since, however, the numbers of 'permitted' states of motion is drastically reduced by the quantum laws, the Pauli-principle plays a very important role in the atomic world: it secures a more or less uniform distribution of electrons around the atomic nucleus and prevents them from crowding in one particular spot.

You must not conclude, however, from the above formulation of the new principle that each of the diffuse quantum-states of motion represented on my diagram may be 'occupied' by one electron only. In fact, quite apart from the motion along its orbit, each electron is also spinning around its own axis, and it will not distress Dr Pauli at all if two electrons move along the same orbit, provided they spin in different directions. Now the study of electron spin indicates that the velocity of their rotation around their own axis is always the same, and that the direction of this axis must always be perpendicular to the plane of the orbit. This leaves only two different possibilities of spinning, which can be characterized as 'clockwise' and 'counter-clockwise'.

Thus the Pauli principle as applied to the quantum states in an atom can be reformulated in the following way: *each quantum state of motion can be 'occupied' by not more than two electrons, in which case the spins of these two particles must be in opposite directions.* Thus, as we proceed along the natural sequence of elements towards the atoms with a larger and larger number of electrons, we find different quantum states of motion being gradually filled with the electrons, and the diameter of the atom steadily increases. It must also be mentioned in this connexion that, from the point of view of the strength of their binding, different quantum states of atomic electrons can be united in separate groups (or shells) of states with approximately equal binding. When we proceed along the natural sequence of elements, one group is filled after another, and, as a consequence of their subsequent filling of electronic shells, the properties of the atoms also change periodically. This is the explanation of the well-known periodic-properties of elements, discovered empirically by the Russian chemist DIMITRIJ MENDELEÉFF.

12

Inside the Nucleus

The next lecture which Mr Tompkins attended was devoted to the interior of the nuclei which make the pivot point for the revolution of atomic electrons.

Ladies and Gentlemen—said the professor—

Digging deeper and deeper into the structure of matter, we will now try to penetrate with our mental eye into the interior of the atomic nucleus, the mysterious region occupying only one thousand billionth part of the total volume of the atom itself. Yet, in spite of the almost incredibly small dimensions of our new field of investigation we shall find it full of very animated activity. In fact, the nucleus is after all the heart of the atom, and, in spite of its relatively small size, contains about 99·97% of total atomic mass.

Entering the nuclear region from the thinly populated electronic atmosphere of the atom, we shall be surprised at once by the extremely overcrowded state of the local population. Whereas electrons of atomic atmosphere move, on the average, distances exceeding by a factor of several hundred thousand their own diameters, the particles living inside the nucleus would literally be rubbing elbows with one another, if only they had elbows. In this sense the picture represented by the nuclear interior is very similar to that of an ordinary liquid, except that instead of molecules we encounter here much smaller and also much more elementary particles known as *protons* and *neutrons*. It may be noticed here that, in spite of having different names, protons and neutrons are now considered simply as two different electric states of the same elementary heavy particle known as the 'nucleon'. Proton is a positively charged nucleon, neutron is an electrically neutral nucleon, and the possibility is not excluded that there are also negative nucleons, although as yet they have never been observed. As far as their geometrical dimensions are concerned, nucleons are not very different from electrons, possessing a diameter of about 0·000,000,000,000,1 cm. But they are much heavier, and a proton or neutron would tip the scales against 1840 electrons. As I have said, the particles forming the atomic nucleus are packed very close together, and this is due to the action of certain special *nuclear cohesive forces*, similar to those acting between the molecules in a liquid. And, just as in liquids, those forces, while preventing the particles from being completely separated, do not hinder their displacement relative to one another. Thus nuclear matter possesses a certain degree of fluidity and, not being disturbed by any external forces, assumes the shape of a spherical drop, just like an ordinary drop of water. In the schematic diagram which I am going to draw for you now, you see different types of nuclei built from protons and neutrons. The simplest is the nucleus of hydrogen which consists of just one proton, whereas the most complicated uranium nucleus consists of 92 protons and 142 neutrons. Of course, you must consider these pictures only as a highly schematic presentation of the actual situation, since, owing to the fundamental uncertainty principle of the quantum theory, the position of each nucleon is actually 'spread out' over the entire nuclear region.

As I have said, particles forming an atomic nucleus are held together by strong cohesive forces, but apart from these attractive forces there are also forces of another kind acting in the opposite direction. In fact, protons, which form about one half of the total nuclear population, carry a positive electric charge, and are consequently repelled from one another by the Coulomb electrostatic forces. For the light nuclei, where the electric charge is comparatively small, this Coulomb repulsion is of no consequence, but in the case of heavier, highly charged nuclei Coulomb forces begin to offer serious competition in the attractive cohesive forces.

When this happens, the nucleus is no longer stable, and is apt to eject some of its constituent parts. That is exactly what happens to a number of elements located at the very end of the periodic system, known as 'radioactive elements'.

From the above considerations you might conclude that these heavy unstable nuclei should emit protons, since neutrons do not carry any electric charge and are therefore not subject to the

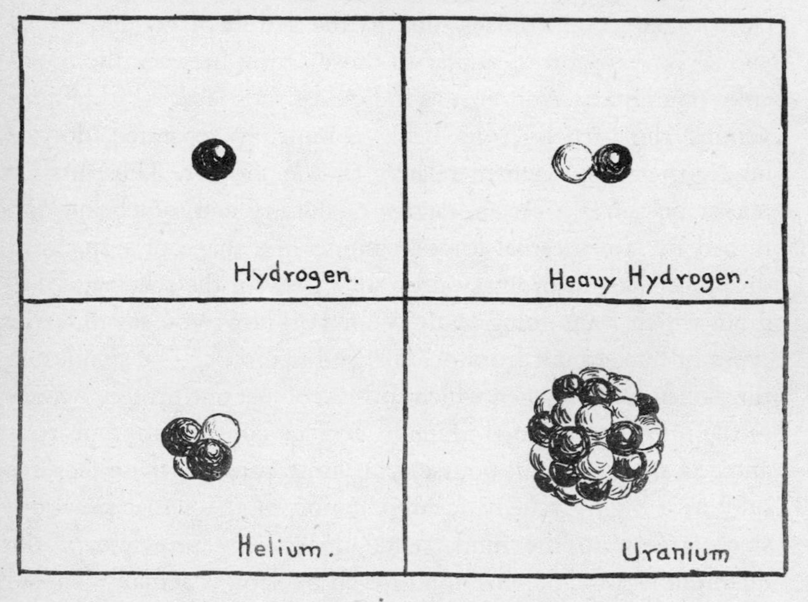

Coulomb repulsive forces. Experiments show us, however, that the particles actually emitted are the so-called *alpha-particles* (helium-nuclei), i.e. complex particles built of two protons and two neutrons each. The explanation of this fact lies in the specific grouping of nuclear constituent parts. It appears that the combination of two protons and two neutrons, forming an alpha-particle, is especially stable, and it is therefore much easier to throw the whole group out at once than to break it into separate protons and neutrons.

As you probably know, the phenomenon of radioactive decay

was first discovered by the French physicist HENRI BECQUEREL, and its interpretation as the result of spontaneous disintegration of atomic nuclei was given by the famous British physicist Lord Rutherford, whose name I have already mentioned before in other connexions, and to whom science owes so great a debt for important discoveries in the physics of the atomic nucleus.

One of the most peculiar features of the process of alpha-decay consists in the sometimes extremely long periods of time needed by alpha-particles in order to make their 'getaway' from the nucleus. For *uranium* and *thorium* this period is measured by billions of years; for *radium* it is about sixteen centuries, and although there are some elements in which decay takes place in a fraction of a second, their life-span can also be considered very long as compared with the rapidity of intra-nuclear motion.

What is it that forces an alpha-particle to stay sometimes for many billions of years inside the nucleus? And if it has already stayed so long why does it finally get out?

To answer this question we must first learn a little more about the comparative strength of the cohesive forces of attraction, and the electrostatic forces of repulsion acting on the particle on its way out of the nucleus. A careful experimental study of these forces was made by Rutherford, who used the so-called 'atomic bombardment' method. In his famous experiments at the Cavendish Laboratory, Rutherford directed a beam of fast moving alpha-particles, emitted by some radioactive substance, and observed the deviations (scattering) of these atomic projectiles resulting from their collisions with the nuclei of the bombarded substance. These experiments confirmed the fact that, while at great distances from the nucleus the projectiles are strongly repelled by electric forces of nuclear charge, this repulsion changes into a strong attraction if the projectile manages to come very close to the outer limits of the nuclear region. You can say that the nucleus is somewhat analogous to a fortress surrounded on all sides by a high, steep bulwark, preventing the particles from getting in as well as from getting out. But the most striking result of Rutherford's experiments consists in the fact that *the alpha-particles getting out of the nucleus in the process of radioactive decay, as well as the projectiles which penetrate into the nucleus from outside, possess actually less energy than would correspond to the top of the bulwark, or the 'potential barrier' as we usually call it.* This was the fact which stood in complete contradiction to all the fundamental ideas of classical mechanics. Indeed, how can you expect a ball to roll over a hill if you have thrown it with far less energy than is necessary to get to the top of the hill? Classical physics could only open its eyes very wide, and suggest that there must have been some mistake in Rutherford's experiments.

But, as a matter of fact, there was no mistake, and if someone was in error it was not Lord Rutherford but classical mechanics itself. The situation was clarified simultaneously by my good friend DR GEORGE GAMOW and by DRS RONALD GURNEY and E. U. CONDON, who pointed out that there is no difficulty whatsoever if one looks at the problem from the point of view of modern quantum theory. In fact, we know that quantum physics today rejects the well defined linear trajectories of classical theory, and replaces them with diffuse ghostly trails. And, just as a good old-fashioned ghost could pass without difficulty through the thick masonry walls of an old castle, these ghostly trajectories can penetrate through potential barriers which seem to be quite impenetrable from the classical point of view.

And please do not think I am joking: the penetrability of potential barriers for particles with insufficient energy comes as a direct mathematical consequence of the fundamental equations of the new quantum mechanics, and represents one of the most important differences between the new and old ideas about motion. But, although the new mechanics permits such unusual effects, it does so only with strong restrictions: in most cases the chances of crossing the barrier are extremely small, and the imprisoned particle must throw itself against the wall an almost incredible number of times before its attempt finally succeeds. The quantum theory gives us exact rules concerning the calculation of the probability of such an escape, and it has been shown that the observed periods of alpha-decay are in complete agreement with the expectation of the theory. Also in the case of projectiles which are shot into the nucleus from the outside, the results of quantum-mechanical calculations are in very close agreement with the experiment.

Before going any further, I want to show you some photographs representing the process of disintegration of various nuclei which were hit by high energy atomic projectiles. (Plate, please!)

In this plate [see p. 142] you see two different disintegration processes photographed in the cloud-chamber which I have described to you in a previous lecture. The picture on the left shows a nitrogen nucleus struck by a fast alpha-particle, and is the first picture of artificial transmutation of elements ever taken. It was made by PATRICK BLACKETT, a pupil of Lord Rutherford. You see a large number of alpha tracks radiating from a powerful alpha-ray source which is now shown in the picture. Most of these particles are passing through the field of vision without a single serious collision, but one of them has just succeeded in hitting a nitrogen nucleus. The track of the alpha-particle stops right there, and you can see two other tracks coming out from the collision point. The long thin track belongs to a proton kicked out from the nitrogen nucleus, whereas the short heavy one represents the recoil of the nucleus itself. This isn't, however, a nitrogen nucleus any more, since by losing a proton and absorbing the incidental alpha-particle it has been transformed into a nucleus of oxygen. Thus we have here an alchemic transformation of nitrogen into oxygen with hydrogen as a by-product.

The second photograph corresponds to nuclear disintegration by the impact of an artificially accelerated proton. A fast beam of protons is being produced in a special high-tension machine, known to the general public as an 'atom-smasher', and enters the

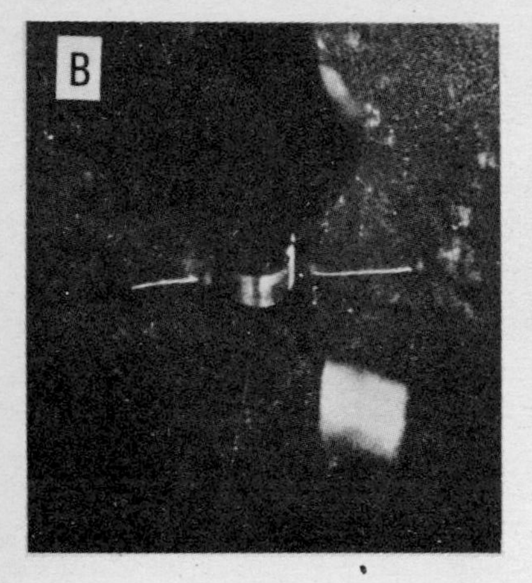

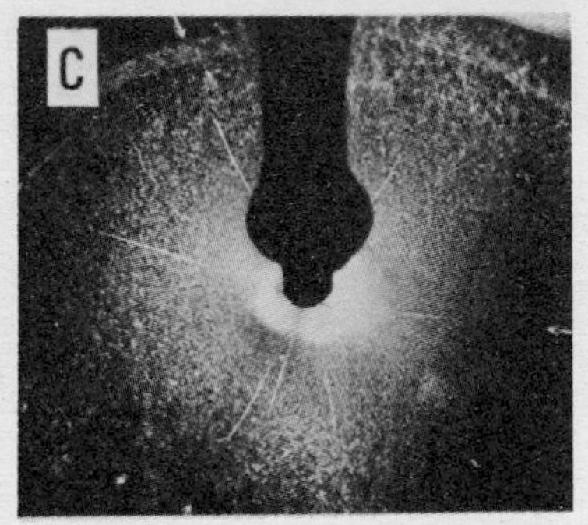

(*a*) Nitrogen hit by helium turns into heavy oxygen and hydrogen

$$_7N^{14} + {_2He^4} \rightarrow {_8O^{17}} + {_1H^1}$$

(*b*) Lithium hit by hydrogen turns into two heliums

$$_3Li^7 + {_1H^1} \rightarrow 2\,{_2He^4}$$

(*c*) Boran hit by hydrogen turns into three heliums

$$_5B^{11} + {_1H^1} \rightarrow 3\,{_2H^4}$$

chamber through a long tube, the end of which is seen in the photograph. The target, in this case a thin layer of boron, is placed at the lower opening of the tube so that nuclear fragments produced in the collision must pass through the air in the chamber, producing cloudy tracks. As you see from the picture, the nucleus of boron, being hit by a proton, breaks into three parts and, counting the balance of the electric charges, we come to the conclusion that each of these fragments is an alpha-particle, i.e. a helium-nucleus. The two transformations shown in the photographs represent rather typical examples of several hundred other nuclear transformations studied in experimental physics today. In all transformations of this kind, known as 'substitutional nuclear reactions', the incidental particle (proton, neutron or alpha-particle), penetrates into the nucleus, kicks some other particle out, and remains itself in its place. We have the substitution of a proton by an alpha-particle, of alpha-particle by proton, proton by neutron, etc. In all such transformations the new element formed in the reaction represents a close neighbour of the bombarded element in the periodic system.

But only comparatively recently, in fact just before the second world war, two German chemists O. HAHN and F. STRASSMANN discovered an entirely new type of nuclear transformation, in which *a heavy nucleus breaks in two equal parts with the liberation of a tremendous amount of energy*. In my next slide (slide please!) [see p. 144] you see on the right a photograph of two uranium fragments flying into the opposite direction from a thin uranium filament. This phenomenon, known as 'nuclear fission', was noticed first in the case of uranium bombarded by a beam of neutrons, but it was soon found that other elements also located near the end of the periodic system possess similar properties. It seems, indeed, that these heavy nuclei are already at the limit of their stability, and that the smallest provocation, caused by a collision with a neutron, is enough to make them break into two, like an oversized drop of mercury. The fact of such instability of heavy nuclei throws light

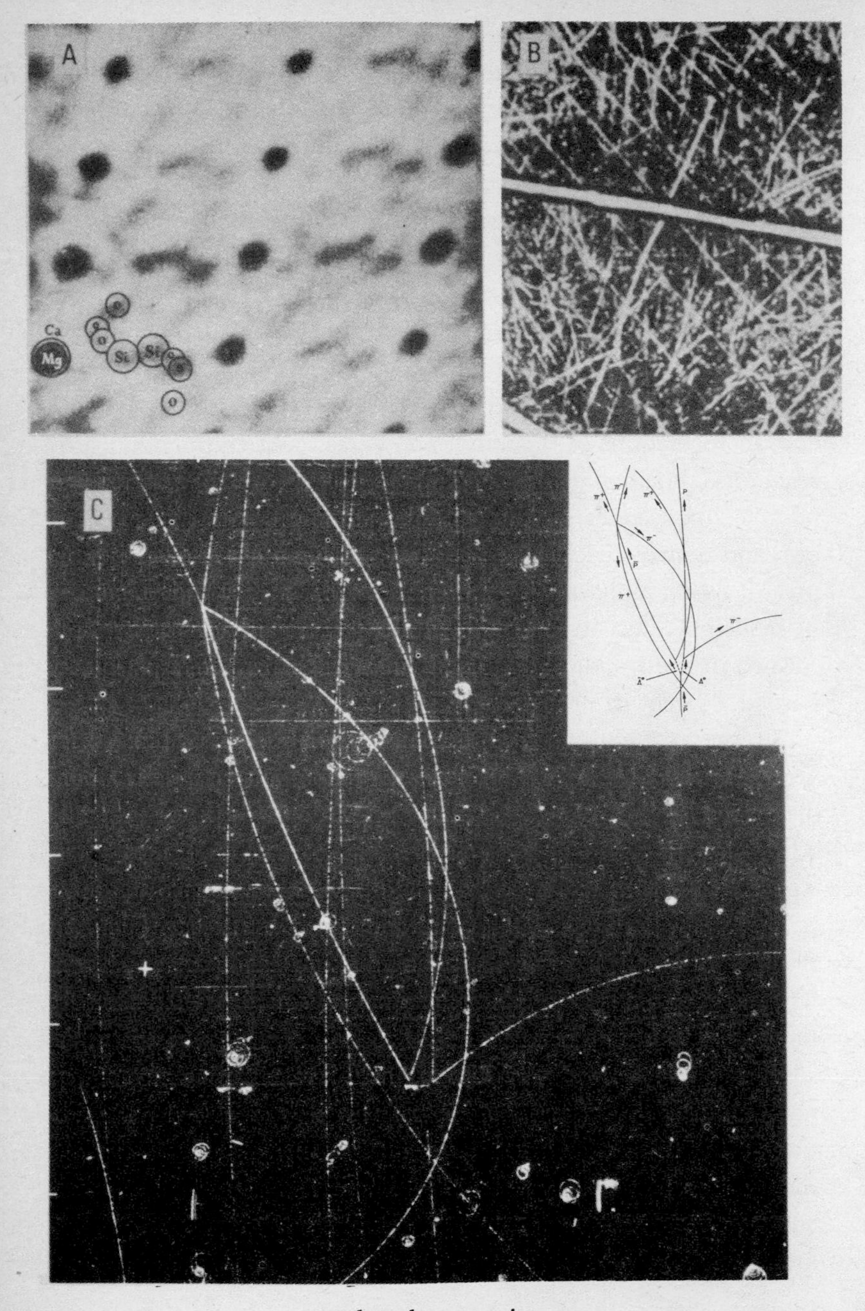

For legend see opposite page

on the question as to why there are only 92 elements in nature; in fact any nucleus heavier than uranium could not exist for any period of time and would immediately break into much smaller fragments. The phenomenon of 'nuclear fission' is also interesting from the practical point of view, since it opens up certain possibilities for the utilization of nuclear energy. The point is that, breaking in half, heavy nuclei also eject a number of neutrons which may cause the fission of neighbouring nuclei. This may lead to an explosive reaction in which all the energy stored inside the nuclei will be set free in a fraction of a second. And, if you remember that the nuclear energy contained in one pound of uranium is equivalent to the energy content of ten tons of coal, you will understand that the possibility of liberating this energy would produce very important changes in our economy.

However, all these nuclear reactions can be obtained only on a very small scale and, although they give us a wealth of information about the internal structure of the nucleus, until comparatively recently there seemed no hope for the release of a vast amount of nuclear energy. It was only in 1939 that the German chemists, O. Hahn and P. Strassmann, discovered an entirely new type of nuclear transformation. In this a heavy nucleus of uranium, hit by a single neutron, breaks into two approximately equal parts, liberating a tremendous amount of energy along with two or three neutrons which, in their turn, may hit other uranium nuclei and break each of them in two also, liberating more energy and more neutrons. This branching fission process may lead to tremendous explosions or, if controlled, supply almost inexhaustible amounts of energy. We are very fortunate that DR TALLERKIN,

---

(*a*) Bragg's photograph of atoms in a diopside crystal. The circles in the corner identify individual atoms of calcium, magnesium, silicon and oxygen. Magnification about 100,000,000

(*b*) Two fission fragments flying in opposite directions from uranium hit by a neutron

(*c*) Production and decay of neutral lambda and anti-lambda hyperons who worked on the atomic bomb and who is also known as the 'father of the hydrogen bomb', agreed to come here in spite of his many commitments and give a short talk on the subject of nuclear bombs. He is due here any minute now.

As the professor said these words the door opened and in came an impressive-looking man with burning eyes and overhanging dark bushy eyebrows. Shaking the professor's hand he turned to the audience.

'*Hölgyeim és Uraim,*' he began. '*Röviden kell beszélnem, mert nagyon sok a dolgom. Ma reggel több megbeszélésem volt a Pentagon-ban és a Fehér Ház-ban. Délutan . . .* Oh, sorry!' he exclaimed, 'sometimes I mix my languages. Let me begin again. Ladies and Gentlemen! I must be short because I am very busy. This morning I attended several conferences in The Pentagon and The White House; this afternoon I have to be present at the underground test explosion at French Flats in Nevada, and in the evening I have to deliver a speech at a banquet at the Vandenberg Air Force Base in California.

'The main point is that atomic nuclei are balanced by two kinds of forces: nuclear attractive forces which tend to hold the nucleus in one piece; and the electric repulsive forces between the protons. In heavy nuclei like those of uranium or plutonium, the latter forces prevail and the nucleus is ready to crack, breaking into two fission products at the slightest provocation. Such a provocation can be provided by a single neutron which hits the nucleus.'

Turning to the blackboard, he continued: 'Here you see a fissionable nucleus and a neutron hitting it. Two fission fragments fly apart, carrying about one million electron volts of energy each and several fresh fission neutrons are also shot out—about two of them, in the case of the light uranium isotope, and about three for plutonium. Then crack! crack! goes the reaction as I have drawn it here on the blackboard. If the piece of fissionable material is small, most of the fission neutrons cross the surface before they have a chance to hit another fissionable nucleus and the chain reaction never starts. But when the piece is larger than what we call a critical mass some three or four inches in diameter, most of the neutrons are trapped and the whole thing blows up. That is what we call a fission bomb, which is often referred to incorrectly as an atomic bomb.

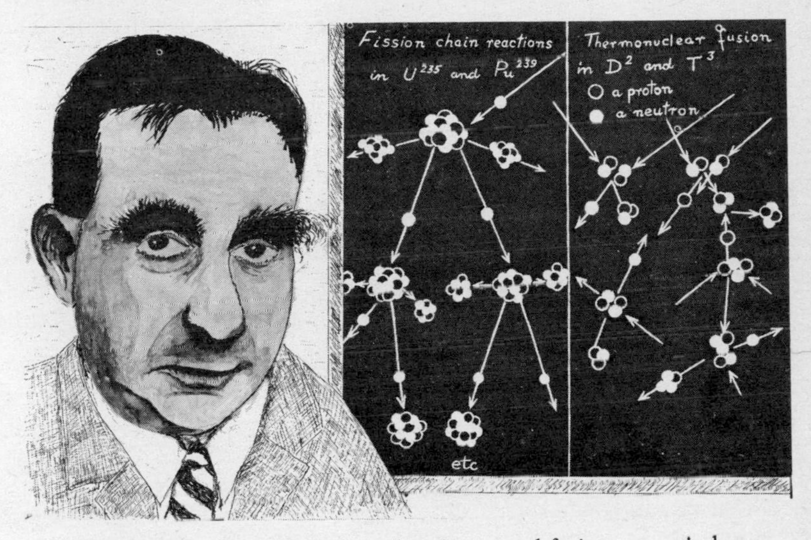

Although the names sound similar, fission and fusion are entirely different processes

'But much better results can be obtained working on the other end of the periodic system of elements where nuclear attractive forces are stronger than electric repulsion. When two light nuclei come into contact they fuse together as do two droplets of mercury on a saucer. This can happen only at a very high temperature, since the light nuclei approaching each other are kept from coming in contact by the electric repulsion. But when the temperature reaches tens of millions of degrees, electric repulsion is impotent to prevent the contact, and the fusion process starts. The most suitable nuclei for the fusion process are deuterons, i.e. the nuclei of heavy hydrogen atoms. There on the right is a simple scheme of thermonuclear reactions in deuterium. When we first thought about the hydrogen bomb, we thought that it would be a blessing to the world, since it produces no radioactive fission products which spread through the atmosphere of the earth. But we were not able to produce such a "pure" hydrogen bomb because deuterium, being the best nuclear fuel which can readily be extracted from ocean water, is still not good enough to burn by itself. Thus we had to surround the deuterium core by a heavy uranium shell. These shells produce a large amount of fission fragments and some people call them "dirty" hydrogen bombs. A similar difficulty is encountered in designing the controlled thermonuclear deuterium reaction, and, in spite of all efforts, we still do not have one. But I am sure this problem will be solved sooner or later.'

'Dr Tallerkin,' asked somebody from the audience, 'what about those fission products from the bomb tests which produce harmful mutations in the population of the entire globe?'

'Not all mutations are harmful,' smiled Dr Tallerkin, 'a few of them are leading to the improvement of progeny. If there were no mutations in living organisms, you and I would still be amoebae. Don't you know that the evolution of life is entirely due to natural mutations and the survival of the fittest?'

'You mean,' shouted a woman in the audience hysterically, 'that we all have to produce children by dozens, and, keeping a few of the best, destroy the rest of them?!'

'Well, Madame,—' started Dr Tallerkin, but at that moment the door of the auditorium opened and in came a man in a pilot suit.

'Hurry, Sir!' he cried, 'your helicopter is parked at the entrance and if we don't start immediately you will miss the connexion with your jetliner at the airport.'

'Sorry,' said Dr Tallerkin to the audience, 'but I must go now. *Isten velük*!' And out they both rushed.

13

The Woodcarver

It was a large and heavy door with an impressive sign, KEEP OUT—HIGH TENSION, right in the middle of it. However, this first inhospitable impression was somewhat softened by the word 'welcome' written large on the door mat, and, after a minute's hesitation, Mr Tompkins pressed the door bell. Let in by a young

assistant, Mr Tompkins found himself in a large room a good half of which was occupied by a very complicated and fantastic looking machine.

'This is our large cyclotron or "atom-smasher", as they call it in the newspapers,' explained the assistant, putting a loving hand on one of the coils of the giant electromagnet which represented the main part of this impressive looking tool of modern physics.

'This is our large cyclotron or "atom-smasher"'

'It produces particles with energy up to ten million electron volts,' he added proudly, 'and there are not many nuclei which can withstand an impact of a projectile moving with such terrific energy!'

'Well,' said Mr Tompkins, 'these nuclei must be pretty tough! Imagine having to build a giant thing like that just to crack the tiny nucleus of a tiny atom. How does this machine work anyway?'

'Have you ever been to the circus?' asked his father-in-law emerging from behind the giant frame of the cyclotron.

'Err...yes, of course,' said Mr Tompkins, rather embarrassed by this unexpected question, 'you mean you want me to go to the circus with you tonight?'

'Not exactly,' smiled the professor, 'but that will help you to understand how a cyclotron works. If you look between the poles of this large magnet, you will notice a circular copper box which serves as a circus ring on which various charged particles, used in experiments on nuclear bombardment, are being accelerated. In the centre of this box is located the source from which these charged particles, or ions, are produced. When they come out, they possess very small velocities, and the strong field of the magnet bends their trajectories into tiny circles around the centre. Then we begin to whip them up to higher and higher velocities.'

'I see how you can whip a horse,' said Mr Tompkins, 'but how you do the same thing with these tiny particles is rather above my head.'

'Nevertheless, it is very simple. If the particle is moving in a circle, all one has to do is to apply to it a series of successive electric shocks each time it passes a given point on its trajectory, just as a trainer in the circus stands on the edge of the ring and whips the horse each time it passes by.'

'But the trainer can see the horse,' protested Mr Tompkins. 'Can *you* see a particle rotating in this copper box to give it a kick just at the proper moment?'

'Of course I can't,' agreed the professor, 'but it isn't necessary. The whole trick of this cyclotron arrangement is that, although the accelerated particle always moves faster and faster, it always executes one complete turn in the same period of time. The point is, you see, that with the increasing velocity of the particle, the radius, and consequently the total length, of its circular trajectory also increases proportionately. Thus it moves along an unwinding spiral, and always comes to the same side of the "ring" at regular intervals. All one has to do is to place there some electric device to give the shocks at regular intervals, and we do it by means of an oscillating electric circuit system, which is very similar to those you can see at any broadcasting station. Each electric shock produced here is not very strong, but their cumulative effect speeds up the particle to extremely high velocities. This is the great advantage of this apparatus; it gives an effect equivalent to that of many million volts, although nowhere in the system are such high tensions actually present.'

'Very ingenious indeed,' said Mr Tompkins thoughtfully. 'Whose invention is it?'

'It was first built by the late ERNEST ORLANDO LAWRENCE at the University of California a number of years ago,' answered the professor. 'Since then cyclotrons have been growing in size and spreading through physical laboratories with the speed of rumour. They seem to be really more convenient than the older devices which used cascade transformers, or machines based on the electrostatic principle.'

'But can't one really break the nucleus without all these complicated devices?' asked Mr Tompkins, who was a great believer in simplicity, and didn't quite trust anything more complicated than a hammer.

'Of course one can. In fact when Rutherford made his first famous experiments on the artificial transformation of elements, he just used ordinary alpha-particles emitted by naturally radioactive bodies. But that was over twenty years ago and, as you can see, the techniques of atom smashing have made considerable progress since then.'

'Can you show me an atom actually being smashed?' asked Mr Tompkins, who always preferred to see things for himself rather than to listen to lengthy explanations.

'Gladly,' said the professor. 'We were just starting an experiment. Here we are making a further study of the disintegration of boron under the impact of fast protons. When the nucleus of a boron atom is hit by a proton hard enough to permit the projectile to pierce the nuclear potential barrier and get inside, it breaks into three equal fragments which all fly in different directions. This process can be directly observed by means of the so-called "cloud chamber" which enables us to see the trajectories of all the particles involved in the collision. Such a chamber, with a piece of boron in the middle, is now attached to the opening of the acceleration chamber, and as soon as we start the cyclotron working you will see the process of nuclear cracking with your own eyes.'

'Will you please switch on the current,' he said, turning to his assistant, 'while I try to tune up the magnetic field.'

It took some time to get the cyclotron started, and left alone Mr Tompkins wandered idly around the lab. His attention was drawn to a complicated system of large amplifier tubes glowing with a faint bluish light. Being quite unaware of the fact that the generating electric tensions used in the cyclotron, though not high enough to crack a nucleus, can easily floor an ox, he leaned forward to look at them more closely.

There was a sharp crack, like that of a lion tamer's whip, and Mr Tompkins felt a terrible shock running through his entire body. The next moment everything went black and he lost consciousness.

When he opened his eyes, he found himself prostrate on the floor where the electric discharge had thrown him. The room around him seemed the same, but all the objects in it had changed considerably. Instead of the towering cyclotron magnet, shining copper connexions, and dozens of complicated electric gadgets attached to every possible spot, Mr Tompkins saw a long wooden work table covered with simple carpenter's tools. On the old-fashioned shelves attached to the wall, he noticed a large number of different wood carvings of strange and unusual shapes. An old, friendly-looking man was working at the table, and, looking more closely at his features, Mr Tompkins was struck by his strong resemblance both to the old man Gepetto in Walt Disney's Pinocchio, and the portrait of the late Lord Rutherford of Nelson hanging on the wall of the professor's lab.

'Excuse my intrusion,' said Mr Tompkins, raising himself from the floor, 'but I was visiting a nuclear laboratory, and something strange seems to have happened to me.'

'Oh, you are interested in nuclei,' said the old man, setting aside the piece of wood he was carving. 'Then you came to just the right place. I make all kinds of nuclei right here and will be glad to show you around my little workshop.'

'You say you *make* them?' said Mr Tompkins rather stupefied.

'Yes, of course. Naturally, it requires some skill, especially in the case of radioactive nuclei, which may fall apart before you even have time to paint them.'

'*Paint* them?'

'Yes, I use red for the positively charged particles and green for the negative ones. Now you probably know that red and green are what one calls "complementary colours", and cancel each other out if mixed together.* This corresponds to the mutual cancellation of positive and negative electric charges. If the nucleus is made up of an equal number of positive and negative charges moving rapidly to and fro, it will be electrically neutral

* The reader must keep in mind that the mixture of colours pertains only to light rays and not to the paints themselves. If we mix red and green paint we shall simply get a dirty colour. On the other hand if we paint one half of a toy top red, and the other green, and then spin it rapidly, it will look white.

and will look white to you. If there are more positive or more negative charges, the whole system will be coloured red or green. Simple, isn't it?'

'Now,' continued the old man, showing Mr Tompkins two large wooden boxes standing near the table, 'this is where I keep the materials from which various nuclei can be built. The first box

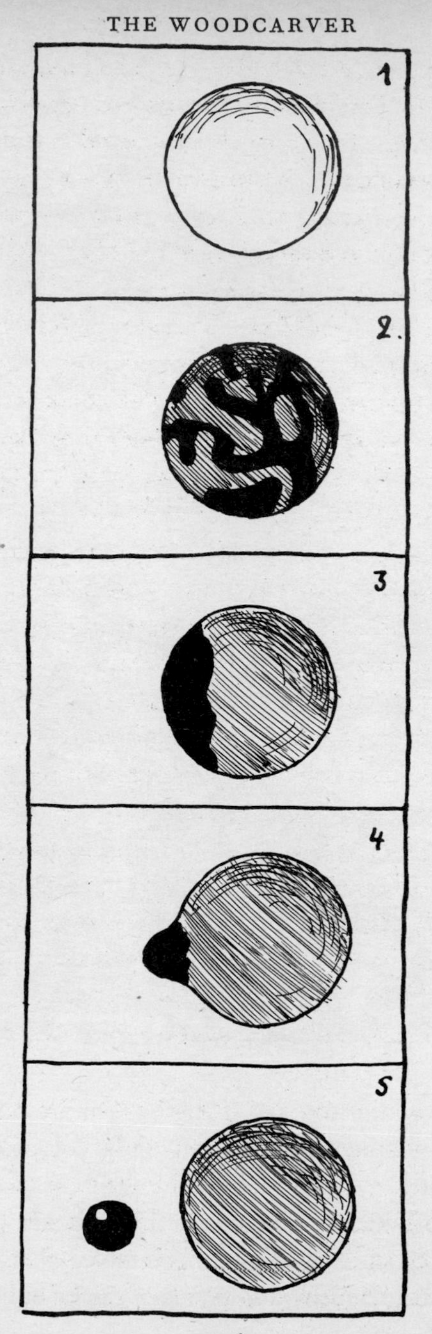
1
2.
3
4
5

contains *protons*, these red balls here. They are quite stable and keep their colour permanently, unless you scratch it off with a knife or something. I have much more trouble with the so-called *neutrons* in the second box. They are normally white, or electrically neutral, but show a strong tendency to turn into red protons. As long as the box is closed tight, everything is all right, but as soon as you take one out, see what happens.'

Opening the box, the old woodcarver took out one of the white balls and placed it on the table. For a while nothing seemed to happen, but just when Mr Tompkins had about lost patience, the ball suddenly came alive. Irregular reddish and greenish stripes appeared on its surface, and for a short while the ball looked like one of the coloured glass marbles children like so much. Then the green colour became concentrated on one side, and finally separated itself entirely from the ball, forming a brilliant green droplet which fell on to the floor. The ball itself was now left completely red, no different from any of the red-coloured protons in the first box.

'You see what happens,' he said, picking the drop of green paint, now quite hard and round, up from the floor. 'The white colour of the neutron broke up into red and green and the whole thing split into two separate particles, a proton and a negative electron.'

'Yes,' he added, seeing the surprised look on Mr Tompkins's face, 'this jade-coloured particle is nothing but an ordinary electron, just like any other electron in any atom or anywhere else.'

'Gosh!' exclaimed Mr Tompkins. 'This certainly tops any coloured handkerchief trick I have ever seen. But can you change the colours back again?'

'Yes, I can rub the green paint back on to the surface of the red ball and make it white again, but that would require some energy, of course. Another way to do it would be to scratch the red paint off, which would take some energy too. Then the paint scratched from the surface of the proton will form a red droplet, that is, a positive electron, about which you have probably heard.'

'Yes, when I was an electron myself...,' began Mr Tompkins, but checked himself quickly. 'I mean, I have heard that positive and negative electrons annihilate each other whenever they meet,' he said. 'Can you do that trick for me too?'

'Oh, it's very simple,' said the old man. 'But I won't take the trouble to scratch the paint off this proton, as I have a couple of positrons left over from my morning's work.'

Opening one of the drawers, he extracted a tiny bright red ball, and, pressing it firmly between finger and thumb, put it beside the green one on the table. There was a sharp noise, like a fire-cracker exploding, and both balls vanished at once.

'You see?' said the woodcarver, blowing on his slightly burned fingers. 'That is why one cannot use electrons for building nuclei. I tried it once, but gave it up right away. Now I use only protons and neutrons.'

'But neutrons are unstable too, aren't they?' asked Mr Tompkins, remembering the recent demonstration.

'When they are alone, yes. But when they are packed tightly in the nucleus, and surrounded by other particles, they become quite stable. However, if there are, relatively speaking, too many neutrons, or too many protons, they can transform themselves, and the extra paint is emitted from the nucleus in the form of negative or positive electrons. Such an adjustment we call a beta-transformation.'

'Do you use any glue, in making the nuclei?' asked Mr Tompkins with interest.

'Don't need any,' answered the old man. 'These particles, you see, stick to each other by themselves as soon as you bring them into contact. You can try it yourself if you want to.'

Following this advice, Mr Tompkins took one proton and one neutron in each hand, and brought them together carefully. At once he felt a strong pull, and looking at the particles he noticed an extremely strange phenomenon. The particles were exchanging colour, becoming alternately red and white. It seemed as if the red paint were 'jumping' from the ball in his right hand to the one in his left hand, and back again. This twinkling of colour was so fast that the two balls seemed to be connected by a pinkish band along which the colouring was oscillating to and fro.

'This is what my theoretical friends call the exchange phenomenon,' said the old master, chuckling at Mr Tompkins' surprise. 'Both balls want to be red, or to have the electric charge, if you want to put it that way, and as they cannot have it simultaneously, they pull it to and fro alternately. Neither wants to give up, and so they stick together until you separate them by force. Now I can show you how simple it is to make any nucleus you want to. What shall it be?'

'Gold,' said Mr Tompkins, remembering the ambition of the medieval alchemists.

'Gold? Let us see,' murmured the old master, turning to a large chart hanging on the wall, 'the nucleus of gold weighs one hundred and ninety-seven units, and carries seventy-nine positive electric charges. That means I have to take seventy-nine protons and add one hundred and eighteen neutrons to get the mass correct.'

Counting off the proper number of particles, he put them into a tall cylindrical vessel and covered it all with a heavy wooden piston. Then, with all his strength, he pushed the piston down.

'I must do this,' he explained to Mr Tompkins, 'because of the strong electric repulsion between the positively charged protons. Once this repulsion is overcome by the pressure of the piston, the protons and the neutrons will stick together because of their mutual exchange forces, and will form the desired nucleus.'

Pressing the piston in as far as it would go, he took it out again and quickly turned the cylindrical vessel upside down. A glittering pinkish ball rolled out on the table, and, watching it more closely, Mr Tompkins noticed that the pinkish colour was due to an interplay of red and white flashes among the rapidly moving particles.

'How beautiful!' he exclaimed. 'So this is an atom of gold!'

'Not an atom yet, only the atomic nucleus,' the old woodcarver corrected him. 'To complete the atom you must add the proper number of electrons to neutralize the positive charge of the nucleus, and make the customary electronic shell around it. But that is easy, and the nucleus itself will catch its electrons as soon as there are some around.'

'Funny,' said Mr Tompkins, 'that my father-in-law never mentioned that one could make gold so simply.'

'Oh your father-in-law and those other so-called nuclear physicists!' exclaimed the old man with a touch of irritation in his voice. 'They put on a fine show but they can actually do very little. They say they cannot compress separate protons into a complex nucleus because they cannot exert great enough pressure to do the job. One of them even calculated that one would need to impose the entire weight of the moon to make the protons stick together. Well, why don't they get the moon if that is their only trouble?'

'But still they produce *some* nuclear transformation,' remarked Mr Tompkins meekly.

'Yes, of course, but awkwardly and to a very limited extent. The quantity of the new elements they get is so small they can hardly see it themselves. I will show you how they do it.' And, taking a proton, he threw it with considerable force against the gold nucleus lying on the table. Nearing the outside of the nucleus, the proton slowed down a little, hesitated a moment and then plunged inside it. Having swallowed the proton, the nucleus shivered for a short time as though in a high fever and then a small part of it broke off with a crack.

'You see,' he said, picking up the fragment, 'this is what they call an alpha-particle, and if you inspect it closely you will notice that it consists of two protons and two neutrons. Such particles are usually ejected from the heavy nuclei of the so-called radioactive elements, but one can also kick them out of ordinary stable nuclei if one hits them hard enough. I must also call your attention to the fact that the larger fragment left on the table is not a gold nucleus any longer; it has lost one positive charge and is now a nucleus of platinum, the preceding element in the periodic table. In some cases, however, the proton which enters the nucleus will not cause it to split in two parts, and as the result you will get the nucleus that follows gold in the table, i.e. the nucleus of mercury. Combining these and similar processes one can actually transform any given element into any other.'

'Oh, now I see why they use fast proton beams produced by the cyclotron,' said Mr Tompkins, beginning to understand. 'But why do you say this method is no good?'

'Because its effectiveness is extremely low. First of all they cannot aim their projectiles the way I can so that only one in several thousand shots actually hits the nucleus. Second, even in the case of a direct hit, the projectile is very likely to bounce off the nucleus instead of penetrating into the interior. You may have noticed when I threw the proton into the gold nucleus that it hesitated somewhat before going in, and I thought for a moment that it was going to be thrown back.'

'What is there to prevent the projectiles from going in?' asked Mr Tompkins with interest.

'You could have guessed it yourself,' said the old man, 'if you had remembered that both the nuclei and the bombarding protons carry positive charges. The repulsive force between these charges forms a kind of barrier which is not so easy to cross. If the bombarding protons manage to penetrate the nuclear fortress, it is only because they use something like the Trojan horse technique; they go through the nuclear walls not as particles but as waves.'

'Well, you have got me there,' said Mr Tompkins sadly, 'I don't understand a word you are saying.'

'I was afraid you wouldn't,' said the woodcarver with a smile. 'To tell you the truth, I'm a workman myself. I can do these things with my hands but I'm not too strong on this theoretical abracadabra either. However the main point is that, as all these nuclear particles are made out of quantum material, they can always go, or rather leak, through obstacles ordinarily considered impenetrable.'

'Oh, I see what you mean!' exclaimed Mr Tompkins. 'I remember that once, shortly before I met Maud, I visited a strange place where billiard-balls behaved exactly the way you describe.

'Billiard-balls? You mean real *ivory* billiard-balls?' repeated the old woodcarver eagerly.

'Yes, I understand they were made from the tusks of quantum elephants,' answered Mr Tompkins.

'Well, such is life,' said the old man sadly. 'They use such expensive materials just for games, and I have to carve protons and neutrons, the basic particles of the entire universe, out of plain quantum oak!'

'But,' he continued, trying to hide his disappointment, 'my poor wooden toys are just as good as all those expensive ivory creations and I will show you how neatly they can pass through any kind of barrier.' And, climbing on the bench, he took from the top shelf a very strange carved figure looking like the model of a volcano.

'What you see here,' he continued, gently brushing off the dust, 'is the model of the barrier of repulsive forces surrounding any atomic nucleus. The outer slopes correspond to the electric repulsion between the charges, and the crater to the cohesion forces which make the nuclear particles stick together. If I now flip a ball up the slope, but not hard enough to bring it over the crest, you would naturally suppose that it would roll back again. But see what actually happens...,' and he gave the ball a slight flip.

'Well, I don't see anything unusual,' said Mr Tompkins, when the ball, after rising about half way up the slope, rolled back again on the table.

'Wait,' said the woodcarver quietly. 'You shouldn't expect it at the first trial,' and he sent the ball up the slope once more. This time it failed again, but at the third attempt the ball suddenly disappeared just when it was about half-way up the slope.

'Well, where do you suppose that one went?' said the old woodcarver triumphantly with the air of a magician.

'You mean it is in the crater now?' asked Mr Tompkins.

'Yes, that is exactly where it is,' said the old man, picking out the ball with his fingers.

'Now, let us get it in reverse,' he suggested, 'and see if the ball can get out of the crater without rolling over the top,' and he threw the ball back into the hole.

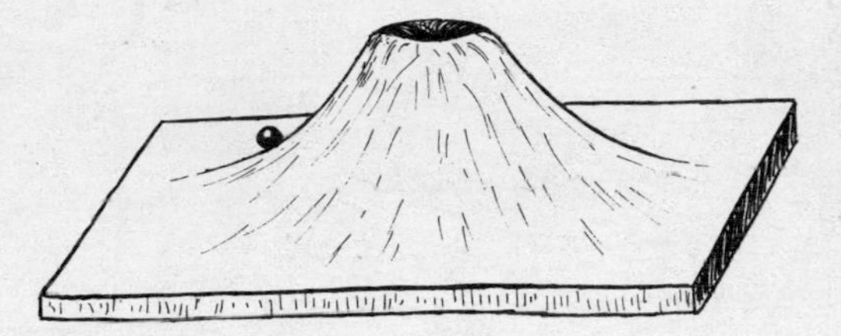

For a while nothing happened, and Mr Tompkins could hear only the slight rumbling of the ball rolling to and fro in the crater. Then, as by a miracle, the ball suddenly appeared in the middle of the outer slope, and quietly rolled down to the table.

'What you see here is a very good representation of what happens in radioactive alpha-decay,' said the woodcarver, putting the model back into its place, 'only there, instead of the ordinary quantum-oak barrier, you have the barrier of repulsive electric force. But in principle there is no difference whatever. Sometimes these electric barriers are so "transparent" that the particle escapes in a small fraction of a second; sometimes they are so "opaque" that it takes many billion years, as for example in the case of the uranium nucleus.'

'But why aren't all nuclei radioactive?' asked Mr Tompkins.

'Because in most nuclei the floor of the crater is below the outer level, and only in the heaviest known nuclei is the floor sufficiently elevated to make such an escape possible.'

It is difficult to say how many hours Mr Tompkins spent in the workshop with the kindly old woodcarver, who was always so eager to communicate his knowledge to anyone who came along. He saw many other unusual things, and above all a carefully closed, but apparently empty casket labelled: NEUTRINOS. *Handle with care and don't let out.*

'Is there something in it?' asked Mr Tompkins, shaking the casket near his ear.

'I don't know,' said the woodcarver. 'Some people say yes, some say no. But you can't see anything anyway. That's a fancy casket given to me by one of my theoretical friends, and I don't quite know what to do with it. Better leave it alone for the time being.'

Continuing his inspection, Mr Tompkins also discovered a dusty old violin, which looked so old that it must have been made by Stradivari's grandfather.

'Do you play the violin?' He turned to the woodcarver.

'Only gamma-ray tunes,' answered the old man. 'It is a quantum-violin, and it doesn't play anything else. Once I had a quantum-cello, for optical tunes, but somebody borrowed it and never brought it back.'

'Well, play me a gamma-ray tune,' asked Mr Tompkins. 'Never heard one before.'

'I will play you "*Nucléet in Th C Sharp*",' said the woodcarver, raising the violin to his shoulder, 'but you must be prepared for it to be a very sad tune.'

The music was very strange indeed, unlike anything Mr Tompkins had ever heard before. There was a steady noise of ocean waves running on sandy shores, interrupted from time to time by a shrill tune reminding him of the whistle of a passing bullet. Mr Tompkins was not exactly musical, but this tune had a weird and powerful effect on him. He stretched himself comfortably in an old armchair and closed his eyes....

14

Holes in Nothing

Ladies and Gentlemen:

Tonight I will request your special attention, since the problems which I am going to discuss are as difficult as they are fascinating. I am going to speak about new particles, known as 'positrons', possessing more than unusual properties. It is very instructive to notice that the existence of this new kind of particle was predicted on the basis of purely theoretical considerations several years before they were actually detected, and that their empirical discovery was largely helped by the theoretical preview of their main properties.

The honour of having made this prediction belongs to a British physicist, Paul Dirac, of whom you have heard and who arrived at his conclusions on the basis of theoretical considerations so strange and fantastic that most physicists refused to believe them for quite a long time. The basic idea of Dirac's theory can be formulated in these simple words: 'There should be holes in empty space.' I see you are surprised; well, so were all physicists when Dirac uttered these significant words. How can there be a hole in an empty space? Does this make any sense? Yes, if one implies that the so-called empty space is actually not so empty as we believe it to be. And, in fact, the main point of Dirac's theory consists in the assumption that *the so-called empty space, or vacuum, is actually thickly populated by an infinite number of ordinary negative electrons packed together in a very regular and uniform way*. It is needless to say that such an old hypothesis did not come to Dirac's mind as the result of sheer fantasy, but that he was more or less forced to it by a number of considerations pertaining to the theory of ordinary negative electrons. In fact, the theory leads to an inevitable conclusion that, besides the quantum states of motion in atoms, there is also an infinite number of special 'negative quantum states' belonging to a pure vacuum, and that, unless one prevents electrons from going over into these 'more comfortable' states of motion, they will all abandon their atoms and will be, so to speak, dissolved into empty space. Since, furthermore, the only way of preventing an electron from going where it pleases, is to have this particular spot 'occupied' by some other electron (remember Pauli), one must have *all* these quantum states in vacuum completely filled up by an infinity of electrons distributed uniformly through the entire space.

I am afraid that my words sound like some kind of scientific abracadabra, and that you cannot make head or tail of all this, but the subject is really very difficult, and I can only hope that if you keep on listening attentively you will be able finally to get some idea about the nature of Dirac's theory.

Well, one way or another, Dirac arrived at the conclusion that empty space is thickly filled with electrons, distributed with a uniform but infinitely high density. How does it happen that we do not notice them at all, and consider the vacuum as an absolutely empty space?

You may understand the answer if you will put yourself in the position of a deepwater fish suspended in the ocean. Does the fish, even if it is intelligent enough to put such a question, realize that it is surrounded by water?

These words brought Mr Tompkins out of a doze into which he had fallen during the beginning of the lecture. He was a bit of a fisherman, and he felt a fresh breeze off the sea and the gently rolling blue waves. But although he was a good swimmer he could not stay on the surface and began to sink deeper and deeper toward the bottom. Strangely enough, he did not feel the lack of air and was quite comfortable. Maybe, he thought, this is the effect of a special recessive mutation.

According to palaeontologists, life originated in the ocean, and the first fishy pioneer to get out on to dry land was the so-called *lungfish* who crawled out to a beach, walking on its fins. According to biologists, these first lungfish, which are called Neoceratodus in Australia, Protopterus in Africa, and Lepidosiren in South America, gradually evolved into land-dwelling animals, like mice, cats, and men. But some of them, like whales and dolphins, after learning of all the troubles of life on dry land, returned to the ocean. Getting back to the water, they retained the qualities acquired during their struggle on the land, and

P. A. M. Dirac was engaged in conversation with a dolphin

remained mammals, the females bearing their progeny inside their bodies instead of just dropping caviar and having it fertilized later by the males. Wasn't it a famous Hungarian scientist named LEO SZILARD* who said that dolphins are more intelligent than human beings?

His thoughts were interrupted by a conversation carried on somewhere deep under the ocean's surface by a dolphin and a typical *homo sapiens* whom Mr Tompkins recognized (from a photograph he had once seen) as the physicist Paul Adrien Maurice Dirac, from Cambridge University.

* Leo Szilard, *The Voice of the Dolphins and Other Stories* (Simon and Schuster, New York, 1961).

'Look here, Paul,' the dolphin was saying, 'you contend that we are not in a vacuum but in a material medium formed by particles with negative mass. As far as I am concerned, water is not different at all from empty space; it is completely uniform and I can move freely through it in all directions. I heard a legend from my pre-pre-pre-pre-predecessor that dry land is quite different, however. There are mountains and canyons which one cannot cross without effort. Here in the water I can move in any direction I choose.'

'You are right in the case of ocean water, my friend,' answered P.A.M. 'Water exerts friction on the surface of your body and if you do not move your tail and fins you will not be able to move at all. Also, because the water pressure changes with the depth, you can float upwards or sink downwards by expanding or contracting your body. But if water had no friction and no pressure gradient you would be as helpless as an astronaut who ran out of rocket fuel. My ocean, which is formed by electrons with negative masses, is completely frictionless and therefore unobservable. Only the *absence* of one of the electrons can be observed by physical instruments, since the absence of a negative electric charge is equivalent to the presence of a positive electric charge, so that even Coulomb could notice it.

'In comparing my electron's ocean with the ordinary ocean we must however make one important exception in order not to be carried too far away by this analogy. The point is that since electrons forming my ocean are subject to the Pauli principle, not a single electron can be added to that ocean when all the possible quantum levels are occupied. Such an extra electron will have to remain above my ocean's surface and can be easily identified by the experimentalists. The electrons discovered first by SIR J. J. THOMSON, the electrons circling around atomic nuclei, or those flying through vacuum tubes, are such excess electrons. And until I published my first paper in 1930 the rest of space was considered to be void, and it was believed that the physical reality belongs only to the occasional splashes rising above the surface of zero energy.'

'But,' said the dolphin, 'if your ocean is unobservable because of its continuity and absence of friction, what is the sense of talking about it?'

'Well,' said P.A.M., 'assume that some external force lifted one of the electrons with negative mass from the depth of the ocean to above its surface. In this case the number of observable electrons will increase by one, which would be considered as a violation of the conservation law. But the empty hole in the ocean from which the electron was removed will now be observable, since the absence of negative charge from a uniform distribution will be perceived as the presence of an equal amount of positive charge. This positively-charged particle will also have a positive mass and will move in the same direction as the force of gravity.'

'You mean it will float up instead of sinking down?' asked the dolphin with surprise.

'Certainly. I am sure you have seen many objects sinking to the bottom, being pulled down by gravity forces: things thrown overboard from ships, or sometimes ships themselves. But look here!' P.A.M. interrupted himself, 'See these small silvery objects rising up to the surface? Their motion is caused by the force of gravity, but they move in the opposite direction.'

'But those are just bubbles,' retorted the dolphin. 'They are probably escaping from something containing the air which turned over or broke, hitting the rocks on the bottom.'

'Right you are, but you would not see bubbles floating up in vacuum. Hence my ocean is not void.'

'Very clever theory,' said the dolphin, 'but is it true?'

'When I proposed it in 1930,' said P.A.M., 'nobody believed it. It was to a large extent my own mistake because I originally suggested that these positively charged particles are nothing else but protons, well known to the experimentalists. You know, of course, that protons are 1840 times heavier than electrons, but I hoped that by some mathematical trick I would be able to explain this increased resistance to acceleration under the action of the given force, and to obtain theoretically the number 1840. But it did not work, and the material mass of the bubbles in my ocean was coming to be exactly equal to that of an ordinary electron. My colleague, Pauli, to whom I must certainly ascribe a sense of humour, was running around professing what he called "the Second Pauli Principle". He calculated, you see, that if an ordinary electron comes close to a hole produced by removal of an electron from my ocean it will fill it up within a negligible period of time. Thus if the proton of a hydrogen atom is really a "hole", it will be instantaneously filled by the ordinary electron rotating around it, and both particles will disappear in a flash of light—or a flash of gamma-rays, should I say. The same would happen, of course, to the atoms of all other elements. Now, the Second Pauli Principle demanded that any theory proposed by a physicist would immediately apply to the matter forming his body, so that I would be annihilated before I had a chance to tell my idea to anybody else. Just like that!' And P.A.M. vanished with a brilliant flash of radiation.

'Sir,' said an irritated voice at Mr Tompkins's ear, 'it is your privilege to sleep at a lecture, but you shouldn't snore. I cannot hear a word of what the professor is saying.'

And, opening his eyes, Mr Tompkins saw again the crowded lecture room and the old professor, who continued:

Let us now see what happens when a travelling hole encounters a surplus electron which is looking for a comfortable place in Dirac's ocean. It is clear that, as the result of such an encounter, the surplus electron will inevitably fall into the hole, filling it up, and the surprised physicist observing the process will register the phenomenon as *the mutual annihilation* of a positive and a negative electron. The energy set free in the fall will be emitted in the form of short-wave radiation, and will represent the only remainder of two electrons who have eaten each other up like the two wolves in the well-known children's story.

But one can also imagine a reverse process in which a pair consisting of a negative and a positive electron are 'created from nothing' by the action of a powerful external radiation. From the point of view of Dirac's theory, such a process consists simply in kicking out an electron from the continuous distribution, and should be considered actually not as a 'creation' but rather as a separation of two opposite electric charges. In the diagram which I now show you, these two processes of electronic 'creation' and 'annihilation' are represented in a very crude schematic way, and

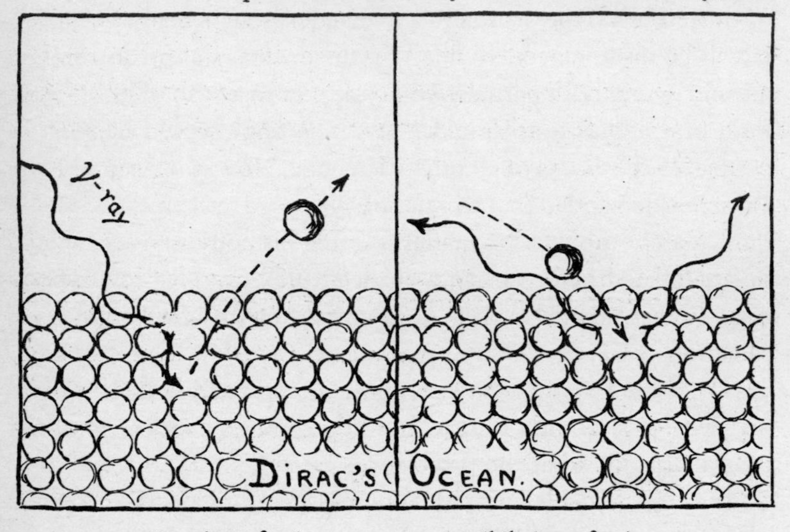

Creation of pair   Annihilation of pair

you see that there is nothing mysterious about the matter. I must add here that, although strictly speaking the process of pair-creation may take place in an absolute vacuum, its probability would be extremely small; you may say that the electron-distribution of a vacuum is too smooth to break it up. On the other hand, in the presence of heavy material particles, which serve as the point of support for the gamma-ray digging into the electronic-distribution, the probability of pair-creation is largely increased and it can be easily observed.

It is clear however that positrons created in the way described above will not exist very long and will soon be annihilated in an encounter with one of the negative electrons which possess large numerical superiority in our corner of the universe. This fact constitutes the reason for the comparatively late discovery of these interesting particles. In fact, the first report on positive electrons was made only in August 1932 (Dirac's theory was published in 1930) by the Californian physicist CARL ANDERSON who, in his studies of cosmic radiation, found particles which resembled in all their aspects ordinary electrons with the only important difference that instead of a negative electric charge they carried a positive one. Soon after this we learned a simple way of producing electron pairs under laboratory conditions by sending a powerful beam of high-frequency radiation (radioactive gamma-rays) through any kind of material substance.

On the next plate I am going to show you, you will see the so-called 'cloud-chamber photographs' of the cosmic-ray positrons, and of the process of pair-creation itself. But before doing so I must explain the way in which these photographs were obtained. The cloud-, or Wilson-chamber, is one of the most useful instruments of modern experimental physics, and it is based on the fact that any electrically-charged particle moving through a gas produces a large number of ions along its track. If the gas is saturated with water vapours, tiny droplets of water will condense on these ions, thus forming a thin layer of fog extending all along the track. Illuminating this foggy band by a strong beam of light on a dark background we obtain perfect pictures, showing all the details of motion.

The first of the two pictures now projected on the screen is the original photograph by Anderson of a cosmic-ray positron, and is, by the way, the first picture of this particle ever taken. The broad horizontal band going across the picture is a thick lead plate placed across the chamber, and the track of the positron is seen as a thin curved scratch going through the plate. The track is curved because during the experiment the cloud-chamber was placed in a strong magnetic field influencing the motion of the particle. The lead plate and magnetic field were employed in order to determine the sign of the electric charge carried by the particle, which can be done on the basis of the following argumentation. It is known that the deflexion of the trajectory produced by the magnetic field depends on the sign of the charge of the moving particle. In this particular case the magnet was placed in such a way that negative electrons would be deflected to the left of the original direction of their motion, whereas positive electrons would be deflected to the

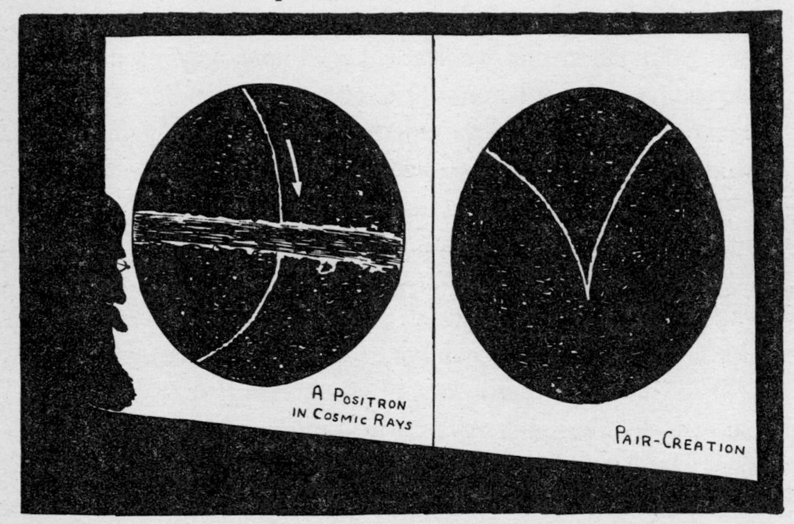

right. Thus if the particle in the photograph was moving upwards it may have had a negative charge. But how to tell which way it was moving? That is where the lead plate comes in. After crossing the plate the particle must have lost some of its original energy, and hence the bending effect of the magnetic field must be larger. In the present photograph the track is bent more strongly *under* the plate (it can hardly be seen at first glance, but comes out in the measurement of the plate). Consequently the particle was moving downwards, and its charge was positive.

The other photograph was taken by JAMES CHADWICK at the University of Cambridge and represents the process of pair creation in the air of the cloud chamber. A strong gamma-ray entering from below, and producing no visible track in the photograph, produced an electronic pair in the middle of the chamber, and the two particles are flying apart, being deflected in opposite directions by the strong magnetic field. Looking at this photograph you may wonder why the positron (which is on the left) is not annihilated on its way through the gas. The answer to this question is also given by Dirac's theory and will be easily understood by anyone who plays golf. If, in putting on the green, you hit the ball too hard, it will not fall into the hole even if your aim is true. In fact a rapidly moving ball will simply jump over the hole and roll on. In the very same way a fast moving electron will not fall into Dirac's hole until its velocity is considerably reduced. Thus a positron has a better chance of being annihilated at the end of its trajectory when it is slowed down by collision along the track. And, as a matter of fact, careful observations show that the radiation which accompanies any annihilation process is actually present at the end of the positron's trajectory. This fact represents an additional confirmation of Dirac's theory.

There remain now two general points still to be discussed. First of all I have been referring to negative electrons as the overflow of Dirac's ocean and to positrons as the holes in it. One can, however, reverse the point of view and consider ordinary electrons as the holes, giving to positrons the role of thrown-out particles. To do this we have only to assume that Dirac's ocean is not overflowing, but that, on the contrary, there is always a shortage of particles. In such a case we can visualize Dirac's distribution to be something like a piece of Swiss cheese with a lot of holes in it. Owing to the general shortage of particles the holes will exist permanently, and if one of the particles is thrown out of distribution it will soon fall back again into one of the holes. It should be stated, however, that both pictures are absolutely equivalent from physical as well as mathematical points of view, and there is actually no difference no matter which one we choose.

The second point can be put in the form of the following question: 'If in the part of the world in which we live there is a definite preponderance in the number of negative electrons, are we to suppose that in some other parts of the Universe this is reversed?' In other words, is the overflow of Dirac's ocean in our neighbourhood compensated for by the lack of these particles somewhere else?

This extremely interesting question is a very hard one to answer. In fact, since atoms built by positive electrons rotating around negative nuclei would have exactly the same optical properties as ordinary atoms, there is no way to decide this question by any spectroscopic observation. For all that we know, it is quite possible that the material forming, let us say, the Great Andromeda Nebula is of this topsy-turvy type, but the only way to prove it would be to get hold of a piece of that material and see whether or not it is annihilated by contact with terrestrial materials. There would be a terrible explosion, of course! There has recently been some talk about the possibility that certain meteorites exploding in the terrestrial atmosphere are formed of this topsy-turvy material, but I don't think that much credit should be given to it. In fact it may very well be that this question of the overflow and draught of Dirac's ocean in different parts of the Universe will remain unanswered forever.

15

Mr Tompkins Tastes a Japanese Meal

One weekend Maud went away to visit her aunt in Yorkshire, and Mr Tompkins invited the professor to have dinner with him in a famous sukiyaki restaurant. Sitting on the soft cushions at a low table, they were enjoying all the delicacies of the Japanese kitchen and sipping sake from little cups.

'Tell me,' said Mr Tompkins. 'The other day I heard Dr Tallerkin saying in his lecture that the protons and the neutrons in a nucleus were held together by some kinds of nuclear forces. Are those the same forces which hold electrons in an atom?'

'Oh, no!' answered the professor. 'Nuclear forces are something quite different. Atomic electrons are attracted to the nucleus by ordinary electrostatic forces first studied in detail by a French physicist, CHARLES AUGUSTIN DE COULOMB, toward the end of the eighteenth century. They are comparatively weak and decrease in inverse proportion to the square of the distance from the centre. Nuclear forces are quite different. When a proton and a neutron come close to each other but not yet in direct contact, there are practically no forces between them. But as soon as they come into contact, there appears an extremely strong force which holds them together. It is like two pieces of adhesive tape which do not attract each other at even a small distance but stick together like brothers as soon as they come in touch with each other. Physicists call these forces 'strong interaction'. They are independent of electric charge of the two particles, and are equally strong between a proton-neutron pair, two protons, or two neutrons.'

'Are there any theories which explain these forces?' asked Mr Tompkins.

'Oh, yes. In the early thirties HIDEKEI YUKAWA proposed

that they are due to the exchange of some as yet unknown particles between the two nucleons; a nucleon is a collective name for a proton and a neutron. When two nucleons come close to each other these mysterious particles begin to jump to and fro between them, leading to a strong binding force which holds them together. Yukawa was able to estimate theoretically their mass, which came to about 200 times larger than the mass of an electron or about 10 times smaller than the mass of a proton or a neutron. Thus they called them "mesatrons". Then the father of Werner Heisenberg, who was a professor of classical languages, objected to this violation of the Greek. The name 'electron', you see, was derived from the Greek ἤλεκτρον meaning *amber*, while "proton" comes from the Greek πρῶτον meaning *first*. But the name of Yukawa's particle is derived from the Greek μέσον meaning *middle*, which has no letter *r* in it. Thus at an international physics meeting Heisenberg proposed to change the name mesatron to "meson". Some French physicists objected because, independent of spelling, meson sounds like *maison*, the French word for home or house. But they were overruled and now the term meson is firmly established. But look at the stage! They are just going to perform a meson show.'

And, indeed, six geishas came out and began to perform a *bilboquet* act in which they were throwing a ball to and fro between two cups which they held in their hands. A man's face appeared in the background singing:

For a meson I received the Nobel Prize,
An achievement I prefer to minimize.
    Lambda zero, Yokohama,
    Eta keon, Fujiyama—
For a meson I received the Nobel Prize.

They proposed to call it *Yukon* in Japan.
I demurred, for I'm a very modest man.
Lambda zero, Yokohama,
Eta keon, Fujiyama—
They proposed to call it *Yukon* in Japan.

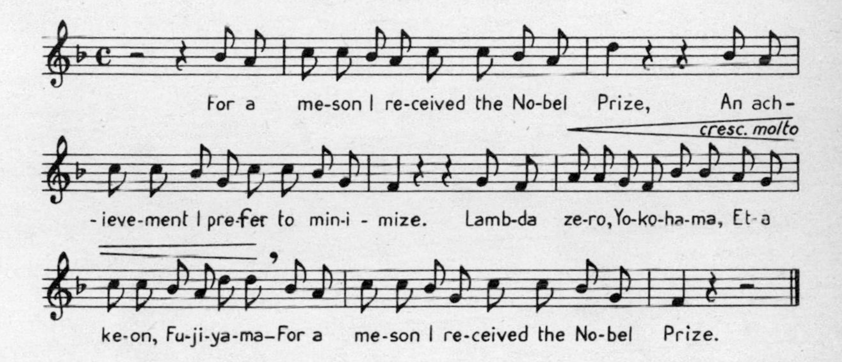

'But why are there three pairs of geishas?' asked Mr Tompkins.

'They represent three possibilities of meson exchange,' said the professor. 'There may be three kinds of mesons: positively charged, negatively charged, and electrically neutral. Maybe all three of them take part in producing nuclear forces.'

'So now there are eight elementary particles,' said Mr Tompkins, counting on his fingers, 'neutrons, protons (positive and negative), negative and positive electrons, and the three kinds of mesons.'

'Ho!' said the professor, 'not eight but closer to eighty. First it was found that there are two kinds of mesons: the heavy and the light mesons designated by the Greek letters π and μ and called *pions* and *muons*. Pions are produced at the fringes of the atmosphere by the impact of the very high energy protons against the nuclei of the gases which form the air. But they are very unstable, and break up, before they reach the Earth's surface, into muons and—most mysterious particle of them all—neutrinos which have neither mass nor charge and are just energy carriers. Muons live somewhat longer, about a few microseconds, so that they manage to reach the earth's surface and decay under our eyes into ordinary electrons and two neutrinos. Then there are also particles designated by the Greek letter κ known as keons.'

Three geishas were playing some unusual *bilboquet* game

'Which kinds of particles did these geishas use in their play?' asked Mr Tompkins.

'Oh, probably pions, the neutral ones, those being the most important, but I am not sure. The majority of new particles which are now being discovered almost every month are so short-lived that, even moving with the speed of light, they decay within the distance of a few centimetres from their origin, so that even the gadgets sent into the atmosphere on balloons do not notice them.

'However, we have now powerful particle accelerators which speed up protons to the same high energy as they reach in cosmic rays: many thousand million electron volts. One of these machines, called the Lawrencetron, is located close by here right up the hill and I will be glad to show it to you.'

A short automobile drive brought them to a large building housing the particle-accelerating machine. Entering the structure, Mr Tompkins was impressed by the complexity of this giant gadget. But, as the professor assured him, it was not more complicated in principle than the slingshot used by David to kill Goliath. The charged particles were entering into the centre of that giant drum, and moving along the unwinding spiral trajectories, being speeded up by alternating electric impulses and kept in line by a strong magnetic field.

'I think I have seen something like that before,' said Mr Tompkins, 'when I visited the Cyclotron, which they used to call an "atom smasher" some years ago.'

'Oh, yes,' said the professor, 'the machine which you have seen before was originally invented by Dr Lawrence. The one you see here is based on the same principle but, instead of accelerating the particles to several million volts, it can speed them to many thousand million volts. Two of them have recently been constructed in the United States. One of them is in Berkeley, California, and is called the Bevatron because it produced particles with the energy of billions of electron volts. It is a strictly American name because in that country a 'billion' is one thousand million. In the United Kingdom a 'billion' means one million million and nobody in good old England has yet tried to achieve that mark. Another American particle accelerator in Brookhaven, Long Island, is called the Cosmotron, which is somewhat overdoing it, because natural cosmic rays often have much higher energy than the Cosmotron can provide. In Europe, at CERN (near Geneva), they have built accelerators comparable to the two American ones. In Russia, not far from Moscow, there is still another machine of that kind, familiarly known as the Khruschevtron, which will probably now be renamed the Brezhnevtron.'

Looking around, Mr Tompkins noticed a door carrying a sign:

ALVAREZ'S LIQUID HYDROGEN
BATHING ESTABLISHMENT

'What is over there?' he asked.

'Oh!' said the professor, 'the Lawrencetron here produces more and more different elementary particles, with higher and higher energies, and one has to analyse them by observing their trajectories and calculating their masses, lifetimes, interactions and

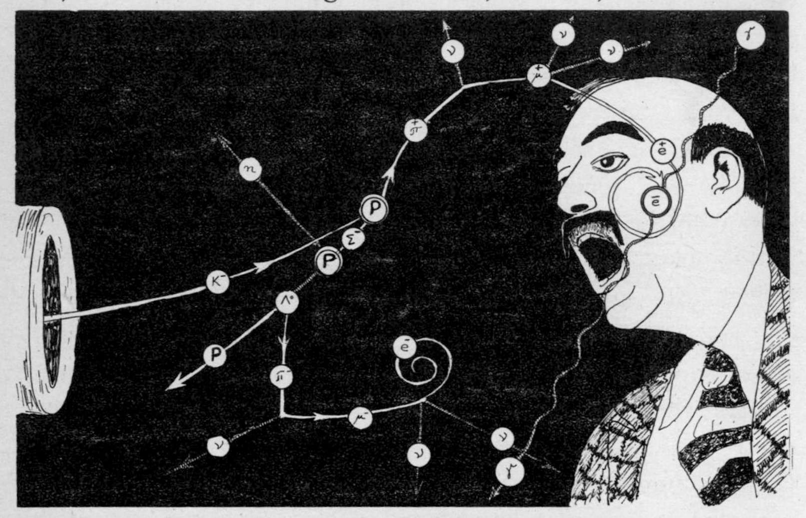

Particles were multiplying like rabbits

many other properties such as strangeness, parity, etc. In olden times one used the so-called cloud chamber invented by C. T. R. WILSON, who received a Nobel Prize for it in 1927. At that time, the fast, electrically-charged particles of a few million electron volts' energy, which were being studied by physicists, were sent through a chamber with a glass top filled with air saturated almost up to the limit by water vapour. When the bottom of the chamber was jerked down, the air in it was cooled by expansion, and the water vapour became *over*saturated. Thus, a fraction of vapour had to condense into tiny water droplets. Wilson discovered that such a condensation of vapour into water goes much faster around ions, i.e. electrically-charged particles of the gas. But gas is ionized along the trajectories of the electrically-charged projectiles passing through the chamber. Thus the foggy stripes of fog, illuminated by a light source located on the side of the chamber, became visible on the black painted bottom of the chamber. You must remember my showing these photographs at the previous lecture.

'Now, in the case of cosmic ray particles with energies a thousandfold larger than those we used to study before, the situation is different because their tracks are so long that the cloud chambers filled with air are too small to follow the tracks from their beginning to their end, and only a small part of the entire picture could be observed.

'A large step forward was made recently by a young American physicist, DONALD A. GLASER, which secured the Nobel Prize for him in 1960. According to his story, he was once sitting gloomily at a bar watching bubbles rising in the beer bottle which stood in front of him. Well, he suddenly thought, if C. T. R. Wilson could study liquid droplets in gas, why can't I do better by studying gas bubbles in liquid? I am not going to discuss technical details,' continued the professor, 'and the difficulties connected with the design of the gadget; it would all be well over your head. But it turned out that, in order to function properly, the liquid in what we now call the bubble chamber had to be liquid hydrogen, the temperature of which is about five and a half hundred degrees Fahrenheit below the freezing point of water. In the next room is a large container built by Louis Alvarez and filled with liquid hydrogen; they usually call it "Alvarez's Bath Tub".'

'Brrrr...it is a bit cold for me!' exclaimed Mr Tompkins.

'Oh, you will not need to get into it. You will just watch the trajectories of the particles through the transparent walls.'

The bathtub was operating as always, and the flash cameras located all around it were taking a continuous row of snapshots.

The bathtub was placed inside of a large electromagnet which was bending the trajectories in order to estimate the speed of their motion.

'It takes only a few minutes to produce a single photograph,' said Alvarez, 'which adds up to several hundred pictures a day, provided that the apparatus does not get out of order and has to be repaired. Each photograph has to be carefully inspected, each track analysed and its curvature carefully measured. It may take anywhere from several minutes to an hour, depending on how interesting the picture is, and how fast the girl analysing it works.'

'Why did you say "girl"?' interrupted Mr Tompkins. 'Is this a purely feminine occupation?'

'Oh no,' said Alvarez, 'many of these girls are actually boys. But in this kind of business we use the term *girl* irrespective of sex, simply as the unit of efficiency and precision. When you say "a typist" or "a secretary" you think about a woman and not a man. Well, to analyse on the spot all the photographs obtained in our laboratory we would need hundreds of girls, which would constitute a problem. Thus we send a large number of our photographs to other universities that do not have enough money to construct the Lawrencetrons and Bubble Baths, but can afford to buy gadgets for analysing our photographs.'

'Are you the only institution doing this job?' inquired Mr Tompkins.

'Oh no! Similar machines exist in Brookhaven National Laboratory on Long Island, New York; in CERN (Corporation Européenne de Recherche Nucléaire) Laboratory near Geneva in Switzerland, and in Shchelkunchik (Nutcracker) Laboratory near Moscow in Russia. They are all looking for a needle in a haystack, and, by God, they find one once in a while!'

'But why is all this work being done?' asked Mr Tompkins in surprise.

'To find new elementary particles, which is more difficult than finding a needle in a haystack, and to study the interaction among

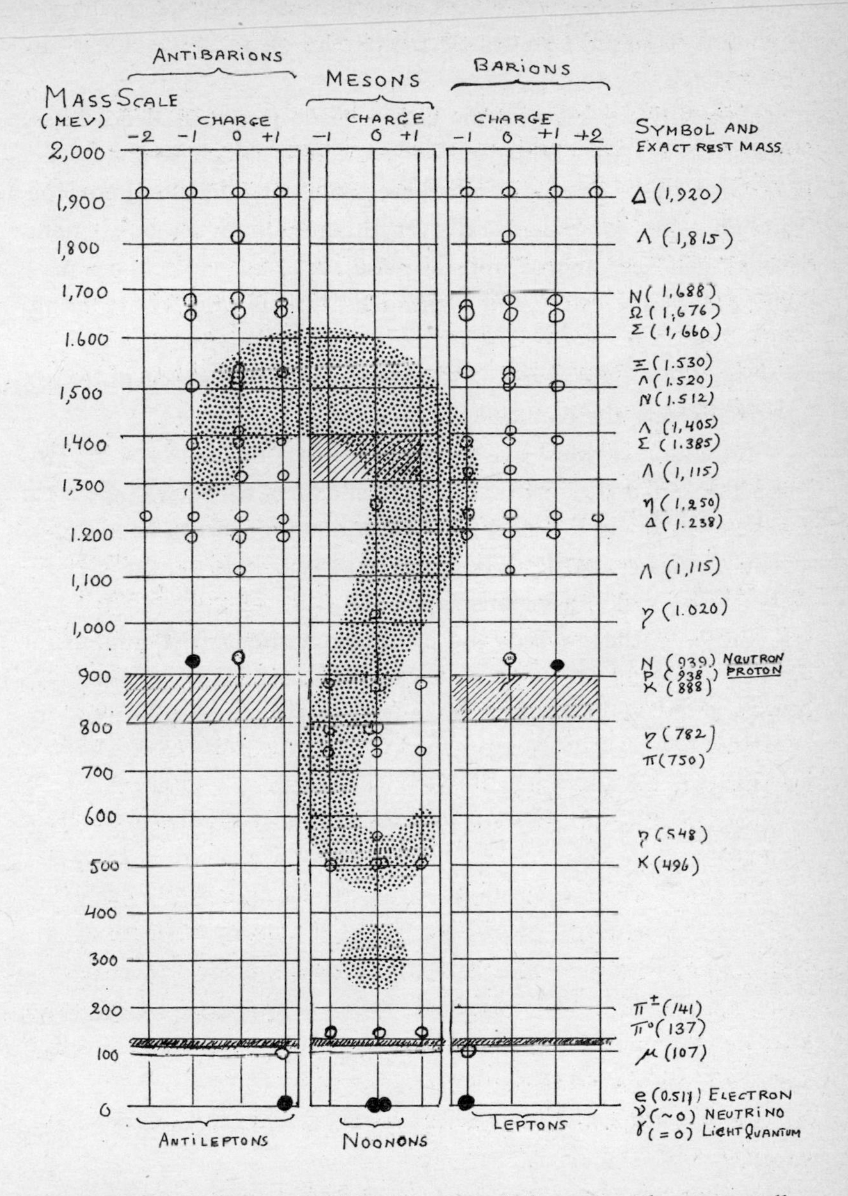

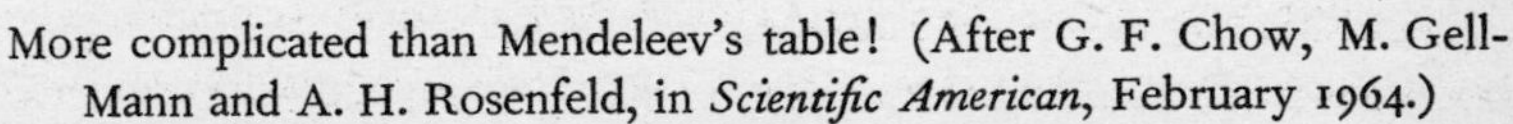

More complicated than Mendeleev's table! (After G. F. Chow, M. Gell-Mann and A. H. Rosenfeld, in *Scientific American*, February 1964.)

them. Here on the wall hangs a chart of particles, and it already contains a larger number of them than there are elements in Mendeleev's system.'

'But why are such terrific efforts made just to find new particles?' asked Mr Tompkins.

'Well, this is science,' replied the professor, 'the attempt of the human mind to understand everything around us, be it giant stellar galaxies, microscopic bacteria, or these elementary particles. It is interesting and exciting and that is why we are doing it.'

'But doesn't the development of science serve practical purposes by improving the comfort and well being of people?'

'Of course it does, but this is only a secondary purpose. Do you think that the main purpose of music is to teach buglers to waken soldiers in the morning, to call them for meals, or to order them to go into battle? They say "curiosity kills the cat"; I say "Curiosity makes a scientist".'

And with these words the professor wished Mr Tompkins a good night.